ELECTRICAL MACHINES & POWER SYSTEMS (Problems with Solutions)

This Page has been intentionally left blank

ELECTRICAL MACHINES & POWER SYSTEMS

(Problems with Solutions)

Dr. C.S. INDULKAR

B. Tech (Hons) (IIT Kharagpur), M.Sc. Tech. (Manchester), Ph.D. (Manchester)

Life Senior Member IEEE, F.I.E. (India)

Associate of the Manchester College of Science & Technology

Former Professor & Head, Electrical Engineering Department (IIT Delhi)

S. CHAND
PUBLISHING

S Chand And Company Limited

(ISO 9001 Certified Company)

S Chand And Company Limited
(ISO 9001 Certified Company)
Head Office: Block B-1, House No. D-1, Ground Floor, Mohan Co-operative Industrial Estate, New Delhi – 110 044 | Phone: 011-66672000
Registered Office: A-27, 2nd Floor, Mohan Co-operative Industrial Estate, New Delhi – 110 044
Phone: 011-49731800
www.schandpublishing.com; e-mail: info@schandpublishing.com

Branches

Ahmedabad	:	Ph: 27542369, 27541965; ahmedabad@schandpublishing.com
Bengaluru	:	Ph: 22354008, 22268048; bangalore@schandpublishing.com
Bhopal	:	Ph: 4274723, 4209587; bhopal@schandpublishing.com
Bhubaneshwar	:	Ph: 2951580; bhubaneshwar@schandpublishing.com
Chennai	:	Ph: 23632120; chennai@schandpublishing.com
Guwahati	:	Ph: 2738811, 2735640; guwahati@schandpublishing.com
Hyderabad	:	Ph: 40186018; hyderabad@schandpublishing.com
Jaipur	:	Ph: 2291317, 2291318; jaipur@schandpublishing.com
Jalandhar	:	Ph: 4645630; jalandhar@schandpublishing.com
Kochi	:	Ph: 2576207, 2576208; cochin@schandpublishing.com
Kolkata	:	Ph: 23357458, 23353914; kolkata@schandpublishing.com
Lucknow	:	Ph: 4003633; lucknow@schandpublishing.com
Mumbai	:	Ph: 25000297; mumbai@schandpublishing.com
Nagpur	:	Ph: 2250230; nagpur@schandpublishing.com
Patna	:	Ph: 2260011; patna@schandpublishing.com
Ranchi	:	Ph: 2361178; ranchi@schandpublishing.com
Sahibabad	:	Ph: 2771238; info@schandpublishing.com

First Edition 2012

ISBN: 978-81-219-3901-0 **Product Code:** H3EMP52ELEC40ENAA12O

PRINTED IN INDIA

By Vikas Publishing House Private Limited, Plot 20/4, Site-IV, Industrial Area Sahibabad, Ghaziabad – 201 010 and Published by S Chand And Company Limited, A-27, 2nd Floor, Mohan Co-operative Industrial Estate, New Delhi – 110 044

PREFACE

This book contains problems in **Electrical Machines & Power Systems** (Problems with solutions). I have used these and other problems in the class - room for many years. In most of the solutions I have deliberately avoided giving theoretical explanations, because an average student should know the theory well before attempting to solve any problem. However, in each chapter, I have provided a brief introduction related to the chapter so that the students are made aware of the contents of the chapter before reading the problems and their solutions. The introduction related to each chpapter contains objective type questions and their answers. The introductions contain brief notes on the topics of the chapters and also include Indian Standards for testing and maintenance of substation equipments, transformers, overhead lines, underground cables and materials. The introductions also contain the http://www addresses of the websites which include electrical machines and power systems topics related to the chapter. After presenting the brief introductory notes on the topic of each chapter, problems with detailed solutions are included in each chapter.

The book contains about 160 objective type questions and an equal number of solved problems. The contents of the book are also available on the author's website: http://cindulkar.tropod.com/index.htm. The solutions to the problems have been developed on Excel spreadsheets. Therefore, educators can use the problems for examination or tutorial purposes, because it is easy to change the problem data on the spreadsheet. Ready availability of the solutions for the new data is an advantage of using the spreadsheet.

To the student, who has this book or who is able to access these problems on the website, my advice is to attempt a problem himself and compare the solution given in this book or the website. If the answer is wrong, he should try to find, from his text-book, where he may have gone wrong and then solve the problem a second time. If he is wrong again, then only he should refer to the solution given in this book or the website. Thus used judiciously, I am sure; the book will be of great help to him. Students preparing for University/Institute of Engineers (India) examinations or the London City & Guilds and other professional examinations will find the book quite useful.

AUTHOR

B3, Gokul Society
Vasna Road
Baroda – 390 015
GUJARAT.
cindulkar@vsnl.net

This Page has been intentionally left blank

CONTENTS

CHAPTER 1

D.C. MACHINES

1.1 INTRODUCTION

Electricity, as is well known, is the most flexible form of energy that can be transported with ease from one place to another for ultimate consumption by its user. In this scenario electrical machines and transformers play a vital role whose importance cannot be overemphasized. Electrical machines are energy converting links between electrical & mechanical networks. Therefore working principles & characteristics of basic machines must be well understood by electrical engineers. In case of DC machines, the following topics must be covered:

- Provision of self and separate excitation.
- Ability of the same machine to work as a generator and a motor.
- Brief reference to the field of application of D.C. Machines.

1.1.1 Constructional features & types of D.C. Machines

Describe by sketches.

1.1.2 Windings

Simple lap & wave windings, their characteristics, winding calculations.

1.1.3 E.m.f and Equations Torque

E.m.f. equation, open circuit characteristics, voltage build -up process for generator, failure of shunt generators to build up, critical resistance.
Torque equation; torque speed characteristics of shunt, series & compound wound motors.

1.1.4 Commutation & Armature Reaction

Commutation: Retarded, accelerated and linear commutation. Methods of improving commutation.

Armature Reaction: shifting of magnetic neutral plane, effect of brush shift, calculation of demagnetizing & cross magnetizing ampere- turns, compensating winding, effect of armature reaction on external characteristics.

Students should be aware of the conventional e.m.f. and torque expressions ($E= \phi\, Z\, n.p/A$) volts and $T = (1/2\pi)\, \phi\, Z\, I\, (p/A)$ N-m.

1.1.5 Characteristics of Series, Shunt, and Compound Wound Generators & Motors

Load characteristics of shunt generators (external & internal), their pre-determination from given open-circuit characteristics and armature resistance drops.

1.1.6 Starting & Speed Control

Starting of D.C. motors, starters.
Basic principles of speed control of a D.C. machine, flux and armature voltage control, speed control of shunt, series & separately excited motors.

1.1.7 Parallel Operation

Parallel operation of shunt and compound wound generators.

1.1.8 Testing & Efficiency

Losses in D.C. machines, efficiency calculation from measurement of losses (Swinburne's method, Hopkinson Test, Field's Test).

1.1.9 Dynamics of D.C. Machines

This Excel spreadsheet module uses a separately excited D.C. motor. The armature inductance is included in the equivalent circuit. The student can study the response of the motor armature current & speed with respect to time for a sudden change in load or applied voltage. The student can vary the parameters like damping (D), armature inductance (L), moment of inertia (J), armature resistance R_a, applied voltage (V), and the load torque (T_L). The student observes the effect of parameter variations on the steady –state & transient response of the system. The response can be obtained either by the Laplace Transform method or by the Runge-Kutta routine.

This module uses a separately excited D.C. motor. The armature inductance is included in the equivalent circuit. The student can study the response of the motor with a sudden change in load or applied voltage. The responses include the variation of armature current and speed with respect to time. For example, the student varies the parameters like damping (D), armature inductance (L), moment of inertia (J), armature resistance (R_a), applied voltage (V), and the load torque (T_L).

Data: Parameters:

$R_a = 0.09724\ \Omega$, $L_a = 10.4$ mH, $J = 0.068$, $K = 1.8$, $D = 0$, $T_L = 0$,
$V = 240$ V, $I_a(0) = 0$, $W_m(0) = 0$, Max. time = 0.2 s.

The student instantly sees the effect of these parameters on the transient and steady state response of the motor. The student can choose either the Laplace transform or the numerical Runge-Kutta algorithm to obtain the response.

THEORY

$$\omega(s)/V(s) = K/[L_a J.s^2 + (L_a . D + R_a J)s + R_a D + K K_b]$$

$$e_b = K_p \frac{d\theta}{dt}$$

$$T = Ki_a$$

Electrical power $e_b.i_a$=mech.power, $T\frac{d\theta}{dt}$

Therefore $K=K_b$

Sudden Change in Applied Voltage, D = 0

$K.V = 432$, $L_a.J = 0.0007$, $R_a.J = 0.0066$, $K^2=3.24$

$$\omega(s) = K.V/s[L_a.J.s^2 + R_a.Js + K^2] = K.E_a(s)[L_a.Js^2 + R_a.Js + K^2] \quad \text{... (1)}$$

$$\omega(s) = 432/s[.000707s^2 + .006612s + 3.24]$$

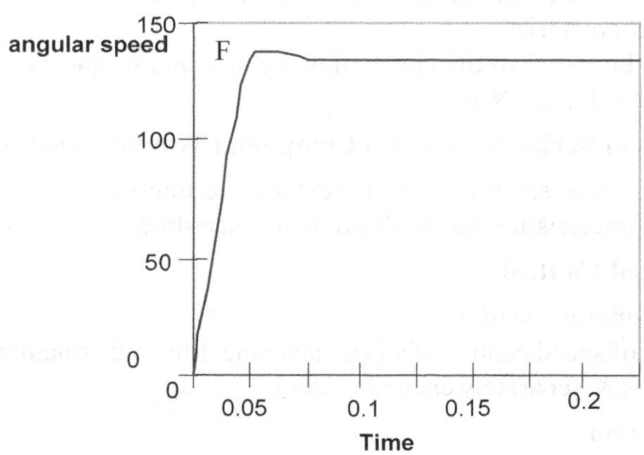

Fig. 1.1(a). *Angular speed Vs. Time*

Angular speed $\omega(t)$ is calculated by solving the above equation using Laplace inverse transform. The result is shown in Fig.1.1.9(*a*).

Armature Current

$I_a(s)/\ \theta(s) = Js^2/K,\ D = 0$

$I_a(s)/\omega(s) = J.s/k$ ... (2)

Product of $w(s)/E_a(s)$ and $I_a(s)/\omega(s)$ gives (from equations (1) & (2) above)

$I_a(s) = (V/L_a)*1/s(s^2 + Ra.s/L_a + K^2/(L_a.J))$

$V/L_a = 0.0231$

$R_a/L_a = 9.35$

$K^2/(La*J) = 4581.4$

$I_a(s) = 023077*1/s(s^2 + 9.35s + 4581.448)$

$I_a\ (t)$ is obtained from the above equation and the response is given in Fig. 1.1.9(*b*).

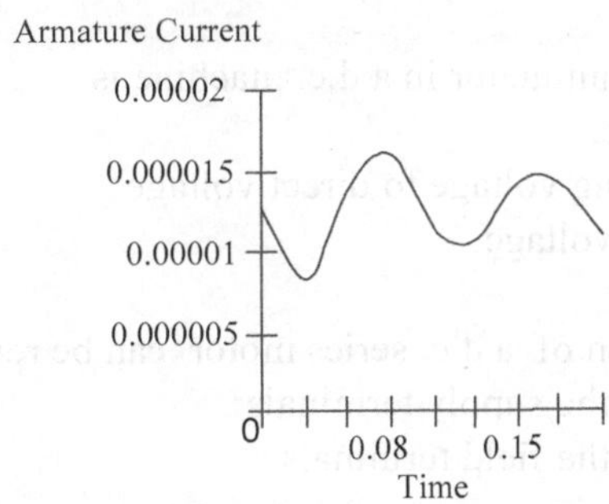

Fig.1.1.9(*b*). *Armature current Vs. Time*

Transient Behaviour

Behaviour under transient conditions is normally approached through the general theory, typical conditions being sudden load or short-circuits on a generator, starting or sudden load change of a motor or a sudden change in voltage applied to a field winding.

Such calculations involve extensive analytical work. Some cases can however lead to a simple numerical solution, as in Prob.1.2.12 below.

Saturation

The theoretical characteristics may have to be significantly modified in practice owing to saturation in the magnetic circuit, especially where the flux changes over a wide range as in a series motor.

The basis for calculations involving the effects of saturation is normally the magnetization curve. The Prob. 1.2.11 below illustrates the procedure for a typical problem.

Other problems in which saturation plays an essential part include the self-excitation of a d.c. generator, effects of high armature reaction and generators with special voltage/current characteristics, e.g., welding generators.

Single - Phase Machines

Of the many types of 1- phase commutator motors that have been devised and used in the past, the only two that have retained commercial importance are the 1-phase series motor and the repulsion motor.

The 1- phase series motor is used on some railways for traction purposes and requires a compensating winding (series connected or short- circuited) to give a satisfactory performance.

Much more common motor is the fractional -kW motor which can run successfully on either a.c. or d.c. and is manufactured on a mass-production basis. Prob.1.2.13 below gives an example of a fractional-kW motor.

OBJECTIVE QUESTIONS

1. The speed of a d.c. series motor at no load is
   (*a*) Zero
   (*b*) Infinity
   (*c*) 1500 r.p.m.
   Ans. (*b*)
2. A sinusoidal voltage of frequency 1 Hz is applied to the field of a d.c. generator. The armature voltage will be
   (*a*) 1 Hz square wave
   (*b*) 1 Hz sine wave
   (*c*) *d.c.* voltage
   Ans. (*b*)
3. The function of the commutator in a d.c. machine is
   (*a*) to change a.c. to d.c.
   (*b*) to change alternating voltage to direct voltage
   (*c*) for easy control of voltage
   Ans. (*b*)
4. The direction of rotation of a d.c. series motor can be reversed
   (*a*) by interchanging the supply terminals
   (*b*) by interchanging the field terminals
   (*c*) either (*a*) or (*b*).
   Ans. (*b*)
5. A d.c. shunt generator driven at normal speed in the normal direction fails to build up armature voltage because
   (*a*) the armature resistance is high
   (*b*) the field current is not sufficiently high
   (*c*) there is no residual magnetism.
   Ans. (*c*)
6. If a d.c. motor is connected across an ac supply, the dc motor will
   (*a*) burn as the eddy currents in the field produce heat
   (*b*) run at its normal speed
   (*c*) run at a lower speed.
   Ans. (*a*)
7. What would happen if the field of dc shunt motors is opened ?
   (*a*) the speed of the motor will be reduced
   (*b*) it will run at its normal speed
   (*c*) the speed will be enormously high, and the motor will destroy itself.
   Ans. (*c*)
8. Which one of the following motors is used in a locomotive ?
   (*a*) Induction motor
   (*b*) dc series motor
   (*c*) Synchronous motor.
   Ans. (*b*)
9. The efficiency of a d.c. shunt generator has a maximum value when.
   (*a*) magnetic losses are equal to mechanical losses

(*b*) armature copper loss is equal to the constant loss
(*c*) field copper loss is equal to the constant loss.
Ans. (*b*)

10. The dummy coil in the d.c. machine is used to
(*a*) eliminate armature reaction
(*b*) bring about mechanical balance of armature
(*c*) eliminate reactance voltage.
Ans. (*b*)

11. The sparking at the brushes in a d.c. generator is attributed to
(*a*) quick reversal of current in the coil under commutation
(*b*) armature reaction
(*c*) reactance voltage
Ans. (*c*)

12. Interpoles in the armature of a d.c. generator are used to
(*a*) neutralize the reactance voltage
(*b*) neutralize the reactance voltage and the cross-magnetizing effect of the armature reaction
(*c*) neutralize the cross-magnetizing effect of the armature reaction.
Ans. (*b*)

13. The mechanical power developed by the d.c. motor is maximum when
(*a*) back e.m.f. = applied voltage
(*b*) back e.m.f. = 0
(*c*) back e.m.f. = 0.5 times the applied voltage.
Ans. (*c*)

14. In series –parallel control method when two d.c. series motors are connected in series, the speed of the set is
(*a*) half of the speed of he motors when connected in parallel
(*b*) ¼ th of the speed of the motors when connected in parallel
(*c*) same as in parallel.
Ans. (*b*)

15. The armature torque of the d.c. shunt motor is proportional to
(*a*) armature current only
(*b*) field flux only
(*c*) armature current and flux both.
Ans. (*c*)

16. The reversal of speed of motors used as electric braking is known as
(*a*) regenerative braking
(*b*) plugging
(*c*) dynamic braking.
Ans. (*b*)

17. The method used for braking in rolling mills is
(*a*) plugging
(*b*) regenerative braking
(*c*) dynamic braking.
Ans. (*a*)

PROBLEMS WITH SOLUTIONS

Prob.1.2.1 D.C. Generator-Back e.m.f.

When a d.c. generator is being driven at 1200 rpm, the generated e.m.f. is 125 V. What will be the generated e.m.f. if (a) the field flux is decreased by 10% with the speed remaining unchanged and (b) if the speed is reduced to 1100 RPM, the field flux remaining unchanged?

Solution. (*a*)

$E_1 = 125$

$\Phi_1 = \Phi_1$

$\Phi_2 = 0.9 \times \Phi_1$

$E_2 = E_1 \times \Phi_2/\Phi_1 = 0.9 \times E_1 = 112.5$ Volts

Answer

(*b*)

$N_1 = 1200$ rpm

$N_2 = 1100$ rpm

$E_2 = E_1 \times N_2/N_1 = 114.583$ Volts **Answer**

Prob.1.2.2 D.C. Generator-generator & motor Action

A shunt machine has armature and field resistances of 0.04 Ω and 100 Ω respectively. When connected to a 460 V d.c. supply and driven as a generator at 600 r.p.m, it delivers 50 kW. Calculate its speed when running as a motor and taking 50 kW from the same supply. Show that the direction of rotation of the machine as a generator and as a motor under these conditions is unchanged.

Solution:

P_1, Generator power = 50000 W

V, Terminal Voltage = 460 V

I, Generator line current = P_1/V = 108.695 A

$R_a = 0.04\ \Omega$

$R_f = 100\ \Omega$

I, Motor line current = 108.695 A

I_{f2} = Motor field current = V/R_f = Generator field current = 4.6 A

I_{a1}, Generator armature current = $I + I_{f2}$ = 113.295

E_1 = Generator back e.m.f. = $V + I_{a1}R_a$ = 464.531 V

I_{a2} = Motor armature current = $I - I_{f2}$ = 104.095 A

E_2, Motor back e.m.f. = $V - I_{a2}R_a$ = 455.836

Speed is proportional to back e.m.f.

N_1, Generator speed = 600 r.p.m.

N_2, Motor speed = $N_1(E_2/E_1)$ = 588.768 r.p.m.

The direction of field flux remains the same when operating as a motor or a generator. However, the direction of the armature current changes in the two modes of operation. Therefore, according to Fleming's Left hand Rule for motor operation and the Right-Hand Rule for the generator operation, the direction of rotation is unchanged.

Prob.1.2.3 D.C. Motor-Output Power & Torque

A 240 V d.c. motor has an armature resistance of 0.68 ohm and draws a a full-load current of 2 4 A at a speed of 100 r.p.m. Calculate (a) the back e.m.f. (b) the output power developed, and (c) the torque developed.

Solution.

V = 240 V

I_a = 24 A

R_a = 0.68 Ω

E, back e.m.f. = $V - I_aR_a$ = 223.68 Volts **Answer**

P, power developed = $E \times I_a$ = 5368.32 Watts **Answer**

N, Speed = 1000 RPM

T, Torque = $\dfrac{60\,P}{2\pi\,N}$ = 51.263 N-m **Answer**

Prob.1.2.4 D.C. Shunt motor-Effects of flux & torque on current and speed

A shunt motor connected across a 440 V supply takes an armature current of 20 A and runs at 500 r.p.m. The armature resistance is 0.6 Ω. If the magnetic flux is reduced by 30 % and the torque developed by the armature increases by 40 %, what are the values of the armature current and of the speed ?

Solution.

V = 440

I_{a1} = 20

R_a = 0.6

$E_1 = V - I_{a1} \times R_a$ = 428 Volts

$\phi_1 = \phi_1$

$\phi_2 = 0.7\,\phi_1$

E_2 = New back e.m.f.

$E_2 = E_1 \times \phi_2/\phi_1 = E_1 \times 0.7$ = 299.6 Volts **Answer**

T_1, Original torque = T_1

T_2, New Torque = 1.4 $T1$

Ia_2 = New Current

N_1 = Original speed 500 r.p.m.

E_1, original back e.m.f. = E_1

T, Torque is proportional to $E \times I_a/N$. Therefore,

$T_2/T_1 = (E_2/E_1)(I_{a2}/I_{a1})(N_1/N_2)$, or

$I_{a2}/N_2 = (T_2/T_1)(I_{a1}/N_1)(E_1/E_2) = 0.08$ . . . (1)

$E_2 = V - I_{a2}R_a$. Therefore,

$I_{a2} = (V - E_2)/R_a$ = 234 Amps

E is proportional to ϕ and N.

Therefore, $N_2 = N_1(E_2/E_1)(\phi_1/\phi_2) = N_1((V - I_{a2}R_a)/E_1)(\phi_1/\phi_2)$

$= 500((440 - I_{a2} \times 0.6)/428)(1/0.7)$

or,

$(440 - I_{a2} \times 0.6)/N_2 = 0.599$ . . . (2)

From equations (1) and (2),

$(440 - I_{a2} \times 0.6)/I_{a2} = 7.49$

or,

I_{a2} = 54.388 A

From Eq .1,

N_2 = 679.851 rpm **Answer**

Prob.1.2.5 D.C. Generator-long Shunt Compound

A 4 pole 500V 25 kW long-shunt compound generator delivers full-load at the rated voltage. Calculate the emf generated, if the armature resistance is 0.03 Ω, series field resistance is 0.04 Ω, and the shunt field resistance 200 Ω. Contact drop per brush is 0.9 volts. Neglect armature reaction.

Solution.

Shunt field

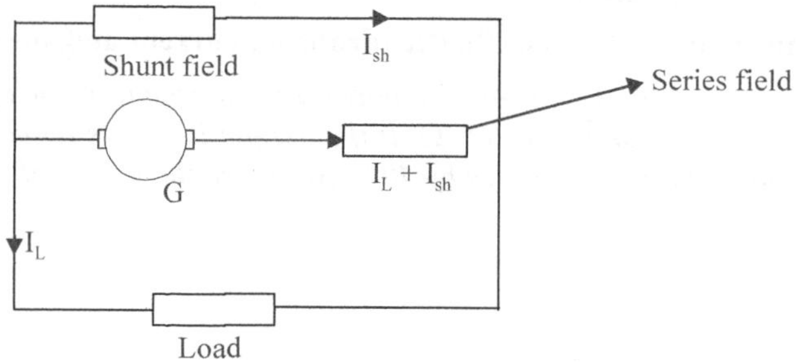

Fig. 1.2

P, Generator power = 25000 W
V, Generator voltage = 500 Volts
I_L, Load current = P/V = 50 A
R_{sh}, Shunt field resistance = 200 Ω
I_{sh} shunt field current = V/R_{sh} = 2.5 A
$I_L + I_{sh}$ = 52.5 A
R_s Series field resistance = 0.04 Ω
Volt drop in series field = $(I_L + Ish) \times R_s$ =2.1 V
R_a Armature resistance = 0.03 Ω
Volt drop in armature = $(I_L + I_{sh})R_a$ =1.575 V
Volt drop in brushes = 2 × 0.9 = 1.8 V
Total voltage drop = 2.1 + 1.575+1.8 = 5.475 V
Terminal voltage = 500 V
E.m.f. = terminal voltage + total voltage drop = 505.475 V **Answer**

Prob.1.2.6 D.C. Generators-parallel operation

Two shunt generators run in parallel on a bus-bar and supply 1000 A to consumers. Each machine has an armature resistance of .02 Ω and a field resistance of 30 Ω, the e.m.f 's generated being 465 V and 460 V respectively. Calculate (a) the output of each machine, and (b) the bus voltage.

Solution. Let

V Bus bar voltage
I_1 Machine 1 current
I_2 Machine 2 current
R_f field resistance = 30 Ω
Total field current = 2 × V/R_f
I_L load current 1000 A
$I_1 + I_2 = I_L + (2 \times V/R_f)$ . . . (1)
R_a = 0.02 Ω
E_1 back emf = 465 Volts
E_2 back emf = 460 Volts
$V = E_1 - I_1 \times R_a = E_2 - I_2 \times R_a$ from which

$I_1 = (E_1 - V)/R_a$ . . . (2)

$I_2 = (E_2 - V)/R_a$ . . . (3)

Substituting Eqs.2 & 3 in Eq. (1)

$((E_1 - V)/R_a) + ((E_2 - V)/R_a) = I_L + (2 \times V/R_f)$

from which,

$V = (E_1 \times R_f + E_2 \times R_f - R_f \times Ra \times I_L)/(2 \times R_a + 2 \times R_f) = 452.198$ Volts

$I_1 = (E_1 - V)/R_a = 640.073$ A

$I_2 = (E_2 - V)/R_a$

$= 390.073$ A

Total current = 640 + 390 = 1030 A = 1000 A of load & 30 A for the fields.

Prob.1.2.7 D.C. Generators in Parallel-effect of changing the excitation of one

Four generators are paralleled on a 240 V bus-bar. All the machines share equally a load of 800 A, the induced emf of each being 250 V. An operator inadvertently alters the position of a field regulator arm thereby increasing the e.m.f. of one machine by 4 %. Calculate the current of each machine and state whether any harm has been done.

Solution. E_1 original e.m.f. of machine1 = 250 V

E_1' new emf of machine1 = $E_1 + .04 \times E'_1 = 260$ V

I_{a1} current of machine1 = 800/4 A

$Ia_1 = 200$ A

V bus bar voltage = 240 V

R_a armature resistance

$I_{a1} R_a = E_1 - V$. Therefore,

$R_a = (E_1 - V)/I_{a1} = 0.05\ \Omega$

V_1 new bus bar voltage

I_{a1}' new current of machine 1

I_{a2}' currents of each of the other three machines

R_L load resistance = 240/800 = 0.3 Ω

$E_1' - R_a \times I_{a1}' = V_1$

$E_2 - R_a \times I_{a2} = V1$

$R_L \times (I_{a1}' + 3 \times I_{a2}) = V1$

Solve the three equations and obtain

$I_{a1}' = 352$ A **Answer**

$I_{a2} = 152$ A **Answer**

$V_1 = 242.4$ Volts **Answer**

Assuming that 200 A is 100% full load current of each machine ,machine1 is loaded more than 175% of its full load capacity. But the air circuit breaker of machine 1 will trip immediately, and the other three machines will be called upon to supply the total load. In short, there will be quite a chaotic condition for a short time in the station, which at all costs must be prevented.

Prob.1.2.8 D.C motor-hysteresis & Eddy Current losses

A d. c motor running at 600 rpm has hysteresis loss of 360 W and eddy current loss of 120 W. Calculate the speed at which the total iron losses will be halved.

Solution. f frequency, k_1, k_2 are constants

W_h, Hysteresis loss = $k_1 \times f$

W_e, Eddy current loss = $k_2 \times f^2$

Since frequency is proportional to speed, N,

$W_h = k_1 \times N$

$W_e = k_2 \times N^2$

$N = 600$ rpm

$W_h = 360\ \Omega$

$W_e = 120\ \Omega$

Therefore,

$k_1 = 0.6$

$k_2 = 0.0003$

N_2, speed at which total iron losses are halved

Half of iron losses = $(W_h + W_e)/2$ =240 W

Therefore ,

$k_1 \times N_2 + k_2 \times N_2 \times N_2 = (W_h + W_e)/2$, a quadratic equation in N_2.

$$N_2 = \frac{-k_1 + \sqrt{k_2{}^2 + 4 \times k_2 (W_h + W_e)/2}}{2 \times k_2} = 337\,\text{rpm}$$ **Answer**

Prob.1.2.9 D.C. Motor-efficiency

A motor is operated at no load from a 240 V supply and the input current is 4 A. If the shunt field resistance is 240 ohm and the armature resistance is 0.32 ohm, calculate

(a) stray loss

(b) efficiency , when input current is 25 A.

Solution.

(a) V, supply voltage = 240 Volts

R_f, field resistance = 240 Ω

I_f, field current = V/R_f = 1A

I, Input current = 4 A

I_a, armature current = 3 A

R_a, armature resistance = 0.32 Ω

$E_g = V - I_a \times Ra$ = 239.04 Volts

W, stray loss = $E_g \times I_a$ =717.12 W **Answer**

(b) I, Input current = 25 A

I_a, armature current = $I - I_f$ = 24 A

E_g, $= V - I_a \times R_a$ = 232.32 Volts

Pout = $E_g \times I_a$-W = 4858.56 Watts

Alternatively,

$P_{out} = P_{in} - P_{loss} = V \times I - I_a{}^2 \times R_a - I_f^2 \times Rf - W$=4858.56 Ω

$P_{in} = V \times I$= 6000 W

Efficiency $= P_{out} \times 100/P_{in}$ = 80.976 % **Answer**

Prob.1.2.10 Universal motor

A fractional kW series motor has a resistance of 20 Ω and an inductance of (1/ π) H. When connected to a 230 V D.C. supply and loaded to take 1 A it runs at 2000 rpm. Determine the speed and power factor when connected to a 230 V, 50 Hz supply and loaded to take the same current.

Solution.

D.C.

V, supply voltage = 230 volts

R, = 20 Ω

I, = 1 A

E, back emf $= V - R \times \mathrm{I} = 210$ volts

N, speed $= 2000$ rpm

On a. c of same value (r.m.s)

L, $= 0.318$ H

f, $= 50$ Hz

X, $= 2 \times 3.14 \times \underline{f} \times L = 99.949$ ohms

$I \times \mathrm{X}$, $= 99.949$ Volts

$I \times \mathrm{R}$, $= 20$ Volts

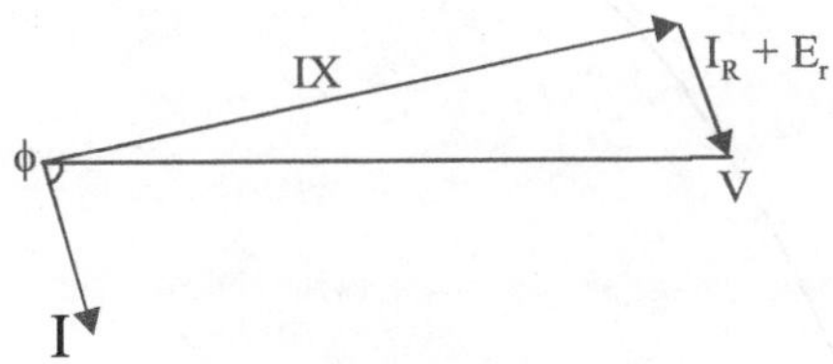

Fig. 1.2.10. *Phasor diagram*

From phasor diagram, reactance drop leads current by 90°, IR drop and rotational emf E_r are in phase with the current (neglecting efffect of iron loss).

Hence ,

E_r = = 207.147 Volts

For the same peak flux, the emf of rotation(rms) is $1/\sqrt{2}$ of D.C. value; but peak flux in a.c is $1\sqrt{2}$ times flux on D.C. for same rms current, therefore E_r is same for same speed and is proportional to speed.

Hence new speed $= E_r \times N/E = 1972.836$ rpm **Answer**

Power factor, $\cos\phi = (IR + E_r)/\mathrm{V} = 0.987$ lag **Answer**

Prob.1.2.11 Field weakening

A d.c series motor having an overall resistance of 0.15 Ω has the following characteristic at 600 V.

Current, A	0	40	80	120	160	200
Speed, rpm	3600	3600	2500	2100	1900	1780

If the field is weakened by tappings which reduce the number of turns/pole to 70% of normal, at what speed would the motor run when taking 160 A from the 600V supply. Neglect armature reaction.

Solution.

The magnetisation characteristic (flux against current at constant speed) is derived from the given torque/current curve as follows :

The emf can be calculated from the resistance and currents. Since the flux is proportional to E/n, the pu flux (taking 200A as producing 1 pu flux) can be found from

$$\phi_{pu} = 1.0(E/E_{200})\ (n_{200}/n)$$

Current, A =	0	40	80	120	160	200
IR drop, V =	0	6	12	18	24	30
E, V =	0	594	588	582	576	570
Φ_{pu} =	0.00	0.52	0.73	0.87	0.95	1.00

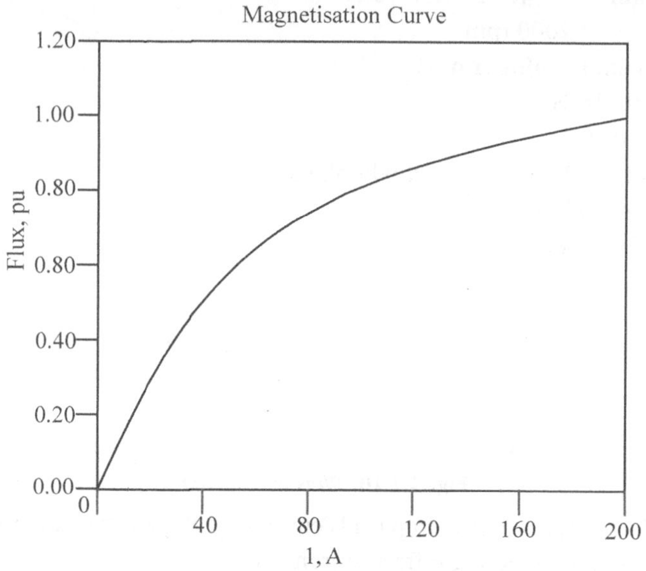

Fig.1.2.11. *Magnetisation curve*

At 160 A, the flux is 0.95. The tappings reduce the effective current to 160 × 0.7= 112 A.

The flux from the above graph at 112 A is about 0.84 pu.

The emf is not altered.

The new speed is = 1900 × 0.95/0.84 = 2149 rpm

The tapping changes the overall motor resistance, but this is neglected.

- Explain why series motors are used in preference to shunt motors for traction applications.
- What would be the effect of armature reaction ?
- Compare the use of diverter resistances and field tappings for series motor speed control.

Prob.1.2.12 D.C generator -Transient behavior

The field winding of a separately excited d. c. generator has an inductance of 75 H, and a resistance of 150 Ω. The rotational mutual inductance coefficient between the field and armature is 1 H/rad/s. With the machine unexcited and driven at 1000 rpm, a ramp function voltage v(t) =100 t volts is applied at t = 0 s to the terminals of the field winding. Determine the armature emf at t = 0.5 s.

Solution:

The voltage equations for the field and armature circuits are:

$v_f = r_f . i_f + L_f di_f/dt = (r_f + L_f p)i_f$

$v_a = w_r . L_{af} i_f$

$v_f = 100t$

Therefore, $v_a = w_r . L_{af} . 100t/(r_f + L_f p)$

Transforming to s-domain for Laplace transforms,

$V_a(s) = \{100/[s.s(s + r_f/L_f)].\}.(w_r .L_{af}/L_f)$

$w_r = 1000 \text{x } 2 \times 3.1416/60 = 104.72$ rad/s.

$r_f = 150$ ohms

$L_f = 75$ H

alpha = $rf/L_f = 2$

$L_{af} = 1$

$\beta = 100\, w_r.L_{af}/L_f = 139.626$

$V_a(s) = \beta/[s^2(s + \alpha)] = \beta[\{(1/s^2))-1/(s(s + \alpha))\}$

$V_a(s) = (1/(\alpha^2)\{(\alpha/s^2)-\alpha/s(s + \alpha))$

Converting to time-domain,

$v_a(t) = [\beta/(\alpha^2)](\alpha.t - 1 + \exp(-\alpha.t))$

with $t = 0.5\ s$

$v_a(t) = 12.841$ sec.

Prob.1.2.13 Fractional kW series motor

A fractional kW series motor has a resistance of 30 Ω and an inductance of 0.5 H. When connected to a 250 V d.c supply and loaded to take 0.8 A it runs at 2000 rpm. Determine the speed and power factor when connected to a 250 V, 50 Hz supply and loaded to take the same current.

Solution. D. C. supply

$I = 0.8$ A

$r = 30$ ohms

$I.r = 24$ Volts

$V = 250$ Volts

$E = V\text{-}Ir$ Volts, Back-emf = 226 volts

A.C supply

$L = 0.5$ H

$X = wL = 157.08\ \Omega$

$IX = 125.664$ volts

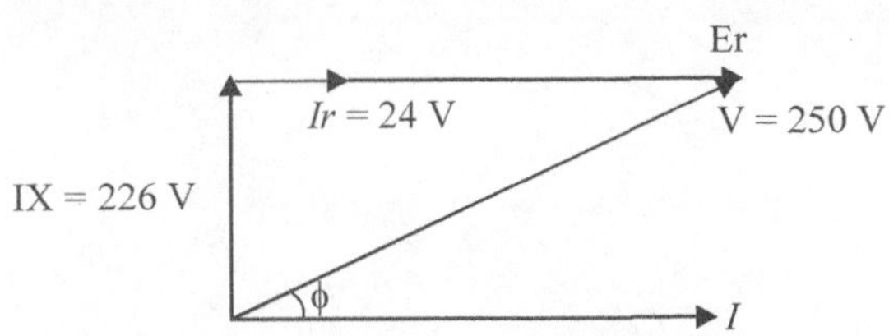

Fig.1.2.13. *Phasor diagram*

Reactance drop leads current by 90 deg; Ir drop and rotational emf E_r are in phase with current (effect of iron loss neglected)

$E_r = \sqrt{v^2 - (I.X)^2} - Ir = 192.122$ volts

For the same peak flux, the e.m.f. of rotation(rad/s) is $\frac{1}{\sqrt{2}}$ of d.c value; but peak flux on a.c is $\sqrt{2}$ times flux on d.c for same r.m.s. current.

Therefore, E_r is same for same speed and is proportional to speed.

Hence, new speed $= 2000 \times E_r/E = 1700.191$ rpm,

p.f $= ((E_r + I_r)/\ V) = 0.86$ lag

Prob. 1.14 Speed Controller of a D.C Motor

A D.C motor is driving a load having an inertia of 200 kgm² and has its speed controlled by a closed-loop system. Frictional torque = 10 Nm/rad/s and the motor torque produced by the error signal is 100 Nm/rad/s of error. If the motor is at rest and the controller is suddenly set to demand a speed of 20 rad/s, develop an expression for the speed in terms of time and determine the steady-state speed which will eventually reach.

Solution. Equation of motion is:

$Jd^2\theta_0/dt^2 + F\, d\theta_0/dt = K\,(d\theta_i/dt - d\theta_0/dt)$

F , Frictional torque

θ_0 , output

θ_i , demanded angular position

Hence,

$d^2\theta_0/dt^2 + (F + K\, d\theta o/dt\,)/Jdt = K\, d\theta i/dt$

Changing to s- domain,

$\theta_0\,(s).s^2 + (((F + K)/J)\theta_0(s).s = (K/(s.J))d\theta i/dt$

$= (1/s)\, d\theta_i / dt$, is a sudden change of input in the form of a step function = 20 rad/s

$(F+K)/J = 0.55$

$K/J = 0.5$

$(K/(s.J))d\theta_i/dt = 10$

Therefore, $\theta_0(s).s^2 + 0.55\, \theta_0(s)\,.s = 10/s$

$\theta_0(s)\quad 10/(s^2(s + 0.55) = (10/0.55)[\,(1/s^2) - (1/0.55)(0.55/(s + 0.55)]$

Taking the inverse transform, we get

$\theta_0(t)\quad (10t/.55\,) \text{-} (10/.55^2\,)\,\left(1-e^{-0.55t}\right)$

Hence speed $d\theta_0/dt = (10/.55) - \dfrac{(10e^{-0.55t})}{0.55^2}$

Steady state speed (at $t = \infty$) = 0/.55 = 18.2 rad/sec.

CHAPTER **2**

SYNCHRONOUS MACHINES

2.1 INTRODUCTION

A Simulator for synchronous generator can be studied on the website *http://powerlearn.ee.iastate.edu.*

2.1.1 Construction & Types of Synchronous Machines

Types of synchronous machines, simple working of synchronous machine as a generator and motor, expression for induced e.m.f. Generator & motor operation.

2.1.2 Armature Reaction

Resultant mmf vector, armature reaction reactance, synchronous impedance, simple equivalent circuit.

2.1.3 Two Axis Theory and Phasor Diagrams

Phasor diagram for machine, synchronous impedance from open circuit & short-circuit tests, short circuit ratio, generator load characteristics, armature reaction ampere turns and leakage reactance, Potier method.

Phasor Diagrams

$V = E - Ir_a - jI_d x_d - jI_q x_q$

where $E = wL_{df} I_f$ and is the emf that would be produced on no load($I_d = I_q = 0$) by the field current I_f (saturation neglected), $wL_d = xd$ (direct-axis synchronous reactance) and $wL_q = x_q$, the quadrature axis synchronous reactance. The phasor diagram of Fig.2.1.3 (*a*) is drawn, for :

(*a*) Salient- pole generator: lagging p.f

(*b*) Non-salient pole generator: lagging p.f ($x_q = x_d = x_s$)

(*c*) Equivalent circuit for (*b*)

(*d*) Salient-pole motor: lagging p.f

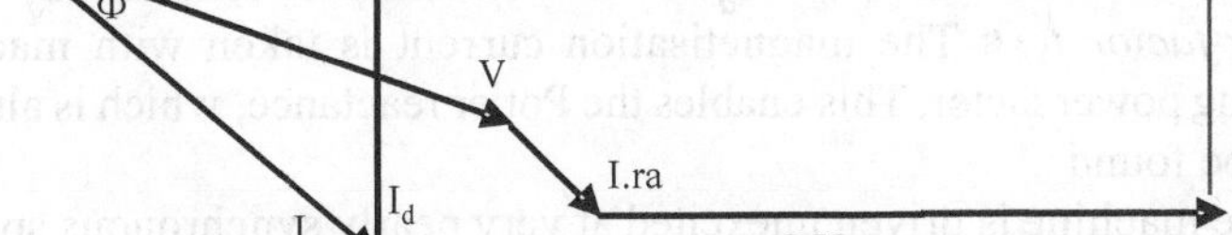

Fig. 2.1.3(*a*) *Salient-pole generator: lagging p.f.*

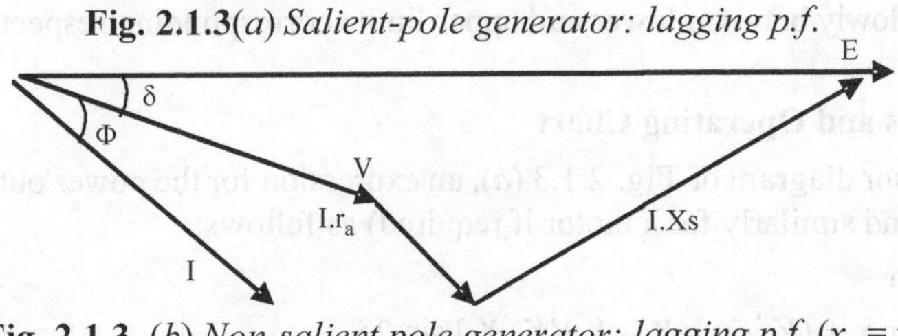

Fig. 2.1.3. (*b*) *Non-salient pole generator: lagging p.f.* ($x_q = x_d = x_s$)

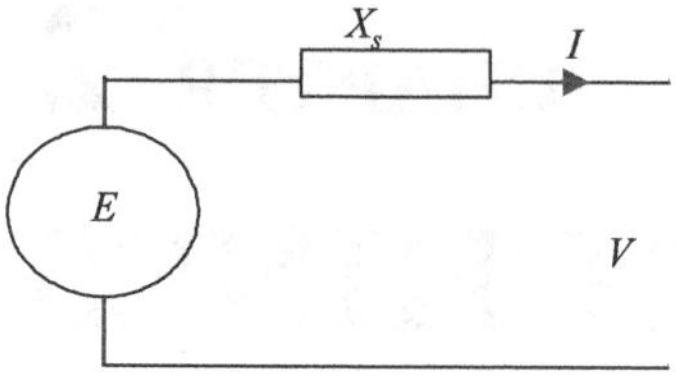

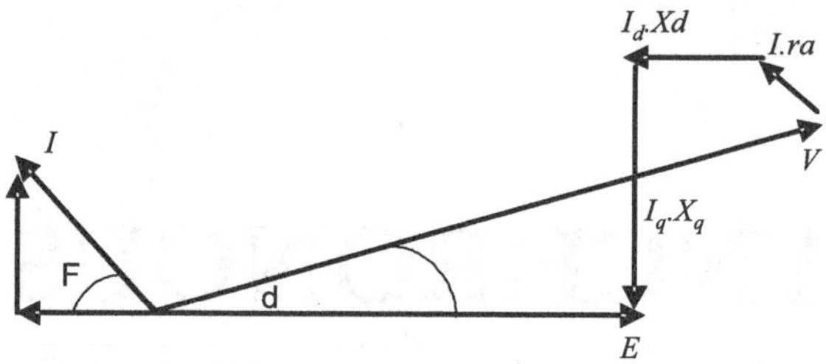

Fig. 2.1.3. (*c*) *Equivalent Circuit for* (*b*) **Fig. 2.1.3.** (*d*) *Salient-pole motor: lagging p.f.*

Fig. 2.1.3. *Synchronous machine diagrams*

For a cylindrical rotor machine (turboalternator), $L_d = L_q$ so that $x_d = x_q = x_s$ and the voltage equation reduces to $V = E - I.x_s$ giving the simpler diagram of Fig.2.1.3 (*b*). The latter diagram leads to a very simple equivalent circuit (per phase) for the machine, as shown in Fig.2.1.3(*c*), resistance usually being negligible. Similar diagrams can be developed for motor operation, the current then being reversed and falling above the horizontal, with corresponding changes in the direction of synchronous reactance voltage drops, as shown in Fig.2.1.3 (*d*).

In the above it has been assumed that the field and armature currents set up separate and independent fluxes. This assumption can, however, only be justified if there is no saturation.

2.1.4 Voltage Regulation

Voltage regulation by different methods

2.1.5 Parallel Operation of Alternators

Synchronous machine on infinite bus (generator & motor action) and synchronization, synchronous machine operation on infinite bus at

1. Constant excitation with variable load
2. Constant load under variable excitation (O-curves & V-curves)

2.1.6 Testing

Students should be familiar with the following four tests. For tests (*i*), (*ii*) and (*iii*) the machine is driven at synchronous speed.

(*i*) *Open-circuit test* (*magnetization curve*): emf plotted against field current.

(*ii*) *Short-circuit test*: Armature current with the armature short-circuited plotted against field current. For a given field current, the voltage from the air line of the magnetization curve divided by the current from the short circuit test gives the direct axis reactance X_d. For a cylindrical rotor machine, the quadrature axis reactance $X_q = X_d$ and for a salient-pole machine $X_q = 0.65$ to $0.7\,X_d$.

(*iii*) *Zero-power-factor test*: The magnetisation current is taken with machine delivering full-load at zero lagging power factor. This enables the Potier reactance, which is almost equal to the leakage reactance to be found

(*iv*) *Slip test*: The machine is driven unexcited at very nearly synchronous speed and with the armature supplied with a low 3-phase voltage. The armature current, and therefore the input impedance, varies slowly between lower and upper limits corresponding respectively, to the values of X_d and X_q.

2.1.7. Power Limits and Operating Chart

From the phasor diagram of Fig. 2.1.3 (*a*), an expression for the power output of a generator can be obtained (and similarly for a motor if required) as follows:

For a generator

$$P = (EV/X_d)\sin\delta + (V^2/2)\,[(X_d - X_q)/X_d.X_q]\sin 2\delta$$

from which a power/angle curve can be drawn. The significance of the second term, which arises from reluctance torque due to the saliency, should be noted.

The phasor diagrams of Fig. 2.1.3(*a*) and (*b*) can be used to develop an operating chart from which the behaviour of the machine under any given conditions can be observed.

The students should be familiar with the effect of varying mechanical input (*e.g.* steam supply) and excitation (field current) when a machine is running on a large network. It is convenient to regard a generator as a device for (*i*) delivering kW and (*ii*) delivering lagging reactive (magnetizing) kVAr to the network, the former being controlled by variation of steam supply and the latter by varying the excitation (field current). The machine requires some field current to produce the necessary magnetization to generate its own emf; if the field current is increased above this value, the excess *I* is available for delivering magnetisation (lagging kVAr) to the system along with the delivery of kW, so that the power factor will be lagging. If the field current is reduced below the self-magnetisation value, magnetizing kVAr must be drawn from the network to make up the deficit, and the power factor will lead.

In the motor, if the field current is less than the self-magnetisation requirement, lagging kVAr and kW will be drawn from the network, and the power factor will lag. Overexcitation results in the generation of an excess of lagging kVAr, and these are sent out to the network although kW are still being received; the power factor is then leading.

The synchronous compensator is often installed at suitable points on a network for the purpose of generating or absorbing lagging kVAr, no kW (except losses) being involved.

2.1.8. Design, Construction and Auxiliaries

(*i*) Design limitations: centrifugal forces and critical speeds, short-circuit ratio, losses and cooling.
(*ii*) Excitation systems: exciter, diode and thyristor equipment, including brushless excitation.
(*iii*) Voltage regulators: Electromechanical and modern electronic and other regulators.

2.1.9 Synchronous Motors

Starting of synchronous motors, V-curves, Torque & Power angle curves.

2.1.10 Motor Starting

The conventional method of starting a synchronous motor is by applying full or reduced voltage to the stator, the motor then running up as a double –cage induction motor with one cage provided by damping windings (or eddy- current paths in the pole face) and the other by the field winding short-circuited through its discharge resistor. The discharge resistor is then disconnected and excitation applied; the motor then pulls into synchronism. The subsequent behaviour depends on the instant in the cycle at which the excitation is applied.

The synchronous –induction motor is a special case of synchronous motor starting, but is not very common in practice.

2.1.11 Synchronising Torque and Hunting

A machine in parallel with others or on infinite bus bars may be subjected to a momentary disturbance causing its rotor to shift by a small angle γ from its normal load angle δ. A synchronizing power P_s and a corresponding torque $T_s = P_s/\omega$ is set up, tending to restore conditions to normal. The synchronising power is

$$P_s = (EV/Z_s) \sin(\theta + \delta) \sin \gamma$$
$$= (EV/x_s)\, \gamma$$

(if the machine is on no-load, γ is small and resistance is neglected)

The torque acts against the rotor inertia J giving (neglecting damping) an equation of motion

$$\frac{Jd^2 g}{dt^2} - T_s\, g = 0$$

This results in simple harmonic motion with a frequency of

$$f = \frac{1}{2\pi}\sqrt{\frac{T_s}{J}}$$

Students can develop the above relations with allowance made for constant (frictional) damping and damping proportional to d γ/dt (due to damper windings).

The practical difficulty of this natural frequency is the possibility of its coinciding with a periodic impressed disturbance of the same frequency *e.g.*, from an IC engine drive, in which case parallel running would be difficult or impossible.

OBJECTIVE QUESTIONS

1. The damper winding in the synchronous motor is often used to
(*a*) prevent hunting only
(*b*) prevent hunting & provide the starting toque
(*c*) maintain synchronism
(*d*) provide the starting torque only.
Ans. (*b*)

2. Which of the following statements is true? Short-pitched coils in the alternators are used to
(*a*) reduce the size of the machine
(*b*) reduce the harmonics or to eliminate the harmonics from the generated emf
(*c*) provide accurate phase difference of 120° between phases
(*d*) reduce copper losses.
Ans. (*b*)

3. The armature flux of an alternator at unity power factor load will be
(*a*) demagnetizing
(*b*) of square waveform
(*c*) cross-magnetizing
Ans. (*c*)

4. The armature reaction of an alternator will be completely magnetizing when load power factor is
(*a*) unity (*b*) zero lagging
(*c*) zero leading (*d*) 0.8 lagging
Ans. (*c*)

5. If two alternators are running in parallel and the voltage of one machine is suddenly increased
(*a*) the machine winding will burn
(*b*) both machines will stop
(*c*) synchronizing torque will be produced to restore synchronism
Ans. (*c*)

6. If the input to the prime mover of an alternator (in a system of three alternators) is kept constant, and the excitation is changed, then
(*a*) the reactive component of the output is changed
(*b*) the active component of the output is changed
(*c*) the power factor of the load remains the same
(*d*) (*a*) and (*b*) occur simultaneously.
Ans. (*a*)

7. The coupling angle or load angle of a synchronous motor is defined as the angle between the
(*a*) rotor and the stator poles of same polarity
(*b*) rotor and the stator poles of opposite polarity
(*c*) rotor and stator teeth
Ans. (*b*)

PROBLEMS WITH SOLUTIONS

Prob.2.2.1 Synchronous generator-Winding distribution factor

Calculate the value of the distribution factor for a 3-phase alternator having 12-slots per pole.

Solution.

slots/pole = 12

α, Electrical angle (radians) between adjacent slots $= \dfrac{180}{12} \times \dfrac{\pi}{180} = 0.262$

m, slots/pole/phase = 4.000

K_d, distribution factor = $\mathrm{Sin}(m\alpha/2)/m.\sin(\alpha/2) = 0.958$ **Answer**

Prob. 2.2.2 Synchronous generator-Induced voltage & distribution factor

A 3-phase 4-pole alternator has a winding with 8 conductors per slot. The armature has a total of 36 slots. Calculate the distribution factor. What is the induced voltage per phase when the alternator is driven at 1800 RPM, with flux of 0.041 Wb in each pole ?

Solution.

ϕ, flux per pole = 0.041 Webers

Z, No. of conductors /phase = $36 \times 8/3 = 96$

N, speed, rpm = 1800

P, No. of poles = 4

f, frequency = NP/120 = 60

slots/pole = 9

$\alpha = (180/9) \times \dfrac{\pi}{180} = \dfrac{\pi}{9} = 0.349$

slots/pole/phase, m = 3

$K_d = \sin\left(\dfrac{m\alpha}{2}\right) \div \sin\left(\dfrac{\alpha}{2}\right) = 0.960$

$V = 2.22\, f\, Z(\Phi)\,(K_d)$ = 503.197 Volts/phase **Answer**

Prob.2.2.3 Synchronous generator-Breadth factor and coil pitch factor

A 10 MVA, 11 KV, 50 Hz, 3-phase star-connected alternator is driven at 300 RPM. The winding is housed in 360 slots and has 6 conductors per slot, the coils spanning (5/6) th of a pole pitch. Calculate the sinusoidally distributed flux per pole required to give a line voltage of 11 kV on open circuit, and the full-load current per conductor.

Solution.

f, frequency = 50 Hz

N, speed, RPM = 300

P = No. of poles = $120f/N = 20$

slots/pole = 360/20 = 18

α, Electrical angle(radians) between adjacent slots = $(180/18) \times (3.1416/180) = 0.175$

m, slots/pole/phase = 6.000

K_d, distribution factor = $\mathrm{Sin}(m\alpha/2)/m\mathrm{Sin}(\alpha/2) = 0.956$

Pitch factor, $K_p = \cos(\beta/2)$

$\beta = 3.1416 - (5/6) \times 3.1416 = 0.524$ radians

$K_p = 0.966$

$E = 2.22 f.Z.\Phi.\, K_d.\, Kp$

E = e.m.f. per phase = 11000/1.73205 = 6350.856 Volts

Z, conductors per phase = 720 (= $360 \times 6/3$)

Φ, flux per pole = $E/2.22 f.\, Z.\, K_d.\, K_p = 0.086$ Webers **Answer**

$I = 10 \times 10E05/(1.73205 \times 11000) = 524.864$ Amps. **Answer**

Prob.2.2.4 Synchronous generator- with unity pf load-Phasor diagram

A 3-phase star-connected 50 Hz alternator has 96 conductors per phase, and a flux per pole of 0.1 Wb. The alternator winding has a synchronous reactance of 5 Ω per phase and negligible resistance. The distribution factor for the stator winding is 0.96. Calculate the terminal voltage when three non-inductive resistors of 10 Ω per phase are connected in star across the terminals. Sketch the phasor diagram.

Solution.

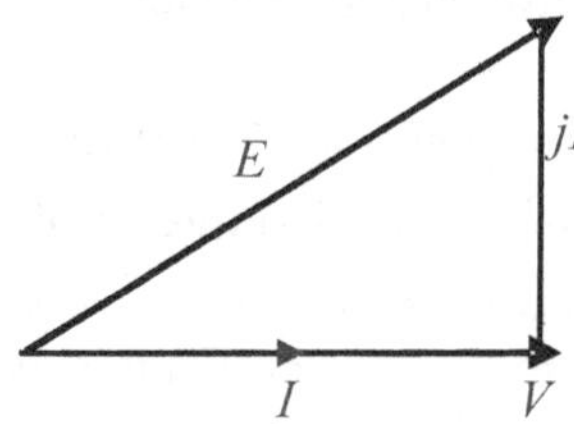

Fig. 2.2.4. *Phasor Diagram*

E = emf per phase

K_d, distribution factor = 0.96

Z, conductors per phase = 96

Φ, flux per pole = 0.1 Wb

f, frequency 50 Hz

E =2.22 f. Z. Φ. K_d = 1022.976 Volts

X, Synchronous reactance = 5 Ω

$V = \sqrt{E^2 - I^2 X^2}$

Load current E = V/load resistance = V/10

Therefore, $V = E = \sqrt{1 + (X^2/10^2)}$ = 914.978 Volts/phase

$V(L-L)$ =1584.787 Volts **Answer**

Prob.2.2.5 Synchronous Generator-Parallel Operation-Example 1

Two single-phase generators are connected in parallel, and the excitation of each machine is such as to generate an open-circuit emf of 3500 V. The stator winding of machine has synchronous reactance of 30 ohms and negligible resistance. If there is a phase displacement of 40 electrical degrees between the emf's ,calculate :

(a) the current circulating between the two machines,

(b) the terminal voltage, and

(c) the power supplied (kW) from one machine to the other.

Assume that there is no external load. Sketch the phasor diagram.

Solution:

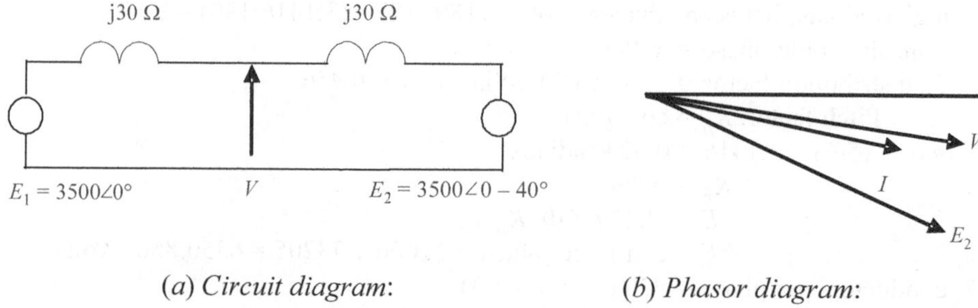

(a) Circuit diagram: *(b) Phasor diagram:*

Fig.2.2.5. *(a) Circuit diagram (b) Phasor diagram*

$I = (E_1 - E_2)/j60 = (3500 - 40)/j60$

$\text{Im } I = (-3500 + 3500\text{Cos}(40 \times 3.14/180))/60 = -13.634$

$\text{Re } I = 3500 \times \text{Sin}(40 \times \frac{\pi}{180°})/60 = 37.480$

$I = \sqrt{\text{ReI}^2 + Im\text{I}^2} = 39.883$ **Answer**

$V = 3500 - j30 \times (\text{ReI} + j\text{Im}I)$

$\text{Re } V = 3500 + 30 \times \text{Im } I = 3090.976$

$\text{Im} V = -30 \times \text{ReI} = -1124.404$

$V = \sqrt{\text{Re}V^2 + ImV^2} = 3289.136$ Volts. **Answer**

Prob.2.2.6 Synchronous generator -Parallel operation-Example 2

Two 500 kVA alternators operate in parallel to supply the following loads:
(a) 250 kW at 0.95 pf lagging
(b) 150 kW at 0.85 pf lagging
(c) 300 kW at 0.75 pf lagging
(d) 100 kW at 0.85 leading

One machine is supplying 400kW at 0.9 pf lagging. Calculate the power factor of the other machine.

Solution.

Load power factors

Cos φ	Sin φ	Load kW	Load kVAr = load kW × Sin φ/Cos φ
0.95 lag	– 0.312	250	– 82.171
0.85 lag	– 0.527	150	–92.962
0.75 lag	– 0.661	300	– 264.575
0.85 lead	0.527	100	61.974
		Total kW = 800	Total kVAr = – 377.733

Machine 1 pf **Machine 1 kW = 400** **Machine1 kVAr = – 193.729**

Cos φ sin φ 400

0.9 lag –0.436 Machine 2 kW = 400

Machine2 KVAr = Total kVAr-Machine 1kVAr = 400 –184.005

Machine 2 kVA = 440.293

Machine 2pf = Machine2 kW / Machine2 kVA = 0.908 lag **Answer**

Prob.2.2.7 Synchronous motor-Power factor improvement

A factory takes 600 kVA at a lagging power factor of 0.6. A synchronous motor is to be installed to raise the power factor to 0.9 lagging when the motor is taking 200 kW. Calculate the corresponding apparent power (in kVA) taken by the motor and the power factor at which it operates.

Solution:

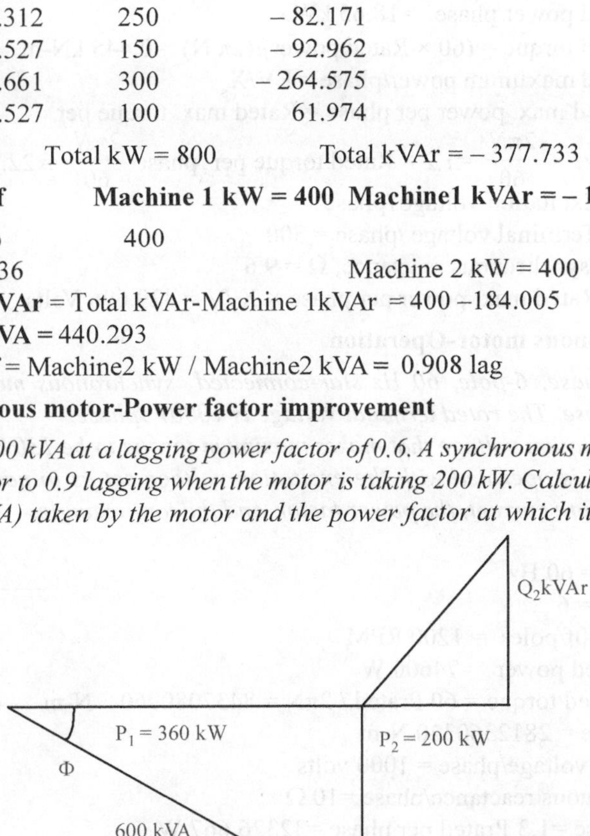

Fig.2.2.7. Phasor diagram

Load power factor, cos Φ = 0.6
Sin Φ 0.8
Load kVA = 600
P_1,Load power = load kVA × cos Φ = 360 kW
Q_1,Load reactive power = kVA × Sin Φ = 480 kVAr
P_2, Motor power = 200 kW
Overall pf, Cos Φ = 0.9 lag
tan Φ = 0.484
Since, tanΦ = $(Q_1 - Q_2)/(P_1 + P_2)$,
$Q_2 = Q_1 - (P_1 + P_2)\tan\Phi$ =208.780 kVAr
S_2, Apparent power of the motor = $\sqrt{P_2^2 + Q_2^2}$ = 289.118 kVA **Answer**

Prob.2.2.8 Synchronous motor Excitation voltage & torque

A 55.95 kW, 3-phase, 6-pole, 60 Hz, star-connected, cylindrical motor has a synchronous reactance of 9.6 Ω/phase. Its rated terminal voltage is 500 V/phase. Find the value of excitation voltage that makes maximum torque to be 120 % of the rated torque.

Solution:

P, No. of poles 6
f, frequency 60
N, speed, rpm = 120 × f/P = 1200
Rated power = 55.95 kW
Rated power/phase = 18.65 kW
Rated torque = (60 × Rated power)/(2π N) = 0.445 kN-m
Rated maximum power/phase = EV/X_s
Rated max. power per phase = Rated max. torque per phase × $\frac{2\pi N}{60}$ =1.2 × Rated torque per phase $\frac{2\pi N}{60}$ = 22.38 kW
E = Excitation voltage /phase
V = Terminal voltage /phase = 500
X_s = synchronous reactance, Ω = 9.6
E = Rated max. power per phase × X_s/V = 429.696 Volts **Answer**

Prob.2.2.9 Synchronous motor-Operation

A 74.6 kW, 3-phase, 6-pole, 60 Hz star-connected synchronous motor has synchronous reactance of 10 Ω/phase. The rated terminal voltage is 1000V/phase.

(a) Find the excitation voltage that makes maximum torque to be 130 % of the rated torque.

(b) The machine is operated with the excitation voltage set as in part (a). For rated load torque, find the armature current, the power factor, and the torque angle.

Solution.

f, frequency = 60 Hz
No. of poles = 6
N, speed =120f/poles = 1200 RPM
Prated = Rated power = 74600 W
T,rated = Rated torque = 60 Prated / 2πN = 8437080960 N-m
T, rated/phase = 2812360320 N-m
V = terminal voltage/phase = 1000 volts
X_s = synchronous reactance/phase =10 Ω
P_{max} per phase =1.3 Prated per phase =32326.667 W
P_{max} = EV/X_s
Therefore, $E = X_s \cdot P_{max}/V$ = 323.267 Volts

P_{rated} per phase = 24866.667 W

$P_{rated} = P_{max} \sin\delta$

$\delta = \sin^{-1}(P_{rated}/P_{max}) = 0.878$ radians

$= 50.285°$

$V = E + j I_a . X_s$,

Therefore $I_a = (V - E)/jX_s = (1000 + j0 - E\cos\delta + j E \sin\delta)/jX_s$

Note: E lags V by δ for a motor

$\text{Re } I_a = E \sin\delta/X_s = 24.867$

$\text{Im } I_a = (-1000+E\cos\delta)/X_s = -71.629$

$I_a = 75.823$ Amps **Answer**

$\cos\Phi = \text{Re } I_a/I_a$ 0.328 lag **Answer**

Prob.2.2.10

Synchronous motor-Operation in parallel with a resistive load

A 3-phase star-connected load takes 50 A current at 0.707 lagging power factor at 220 V between the lines. A 3-phase star-connected round rotor synchronous motor, having a synchronous reactance of 1.27 Ω / phase is connected in parallel with the load. The power developed by the motor is 33 kW at a power angle of 30°. Neglecting armature resistance ,calculate

(a) reactive kVA of the motor, and

(b) the overall power factor of the motor and the load.

Solution.

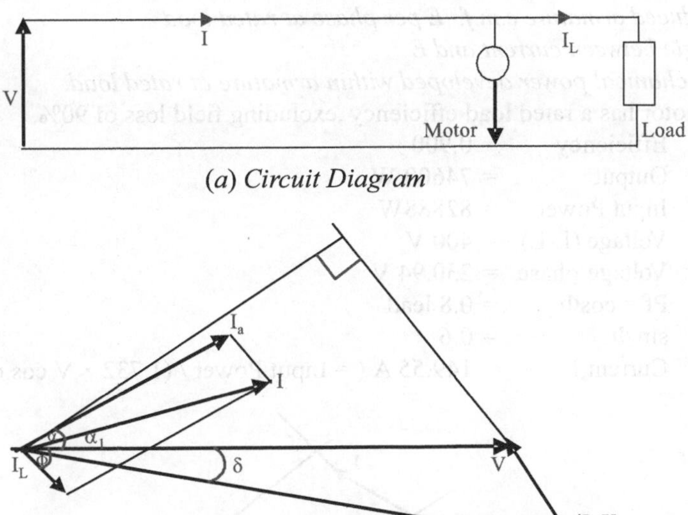

Fig.2.2.10

$\cos\phi = 0.707$

$\sin\phi = 0.707$

I_L = load current = 50 A

V(L-L) 220.000 Volts

V per phase = 127.017 Volts

P, motor Power/phase = 11000 W

$X_s = 1.270\ \Omega$

δ = load angle, (deg) = 30°
radians = 0.524
$P = VE \sin\delta / X_s$
Therefore, $E = P X_s / V \sin\delta$ = 219.970 Volts
$E \cos\delta = V + I_a X_s \cos\alpha$
$E \sin\delta = I_a X_s \sin\alpha$ Therefore,
$I_a^2 \times X_s^2 = E^2 + V^2 - 2EV\cos\delta$
Substituting for X_s, E, V, $\cos\delta$, we get
I_a = 99.993 Amps **Answer**
$\cos\alpha$ = motor pf = $P/(V\ I_a)$ = 0.866 lead **Answer**
$\sin\alpha$ = 0.500
Motor kVAr = $1.73 \times V\ I_a\ \text{Sin}(\alpha)$ = 19047.226 Var **Answer**
Overall power factor angle is given by $\tan\alpha_1 = (I_a.\sin\alpha - I_L \sin\phi)/(I_a \cos\alpha + I_L \cos\phi)$
$\therefore \tan\alpha_1 = 0.120$
Overall pf = 0.993 lead **Answer**

Prob.2.2.11

Synchronous motor-Induced e.m.f. & Mechanical power developed

A 74.6 kW 400 V 1200 RPM, 3-phase Y-connected synchronous motor has an armature resistance of 0.05 Ω/phase and a leakage reactance of 0.5 Ω / phase. At rated load and 0.8 pf leading current, determine,

(*a*) *induced armature e.m.f. E per phase at rated load*
(*b*)*angle between current and E*
(*c*) *mechanical power developed within armature at rated load.*

The motor has a rated load efficiency ,excluding field loss of 90%.

Solution;

Efficiency	= 0.900
Output	= 74600 W
Input Power	= 82888W
Voltage (L-L)	= 400 V
Voltage/phase	= 230.94 V
Pf = $\cos\Phi$	= 0.8 lead
$\sin\Phi$	= 0.6
Current,I	= 149.55 A (= Input Power / (1.732 × V cos ϕ)

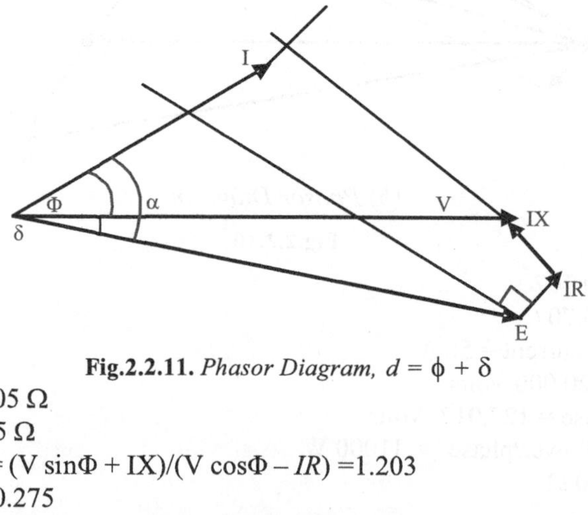

Fig.2.2.11. *Phasor Diagram, d* = $\phi + \delta$

R = 0.05 Ω
X = 0.5 Ω
$\tan\alpha = (V \sin\Phi + IX)/(V \cos\Phi - IR)$ = 1.203
$\alpha°$ = 50.275 **Answer**
$E = V - IZ = V - (I\cos\Phi + jI\sin\Phi)(R + jX)$

ReE = V – IR cosΦ + IX sinΦ =269.823

ImE = – IXcosΦ + IRsinΦ = – 55.333

E = 275.438 Volts **Answer**

Mechanical, Power =3 EIcosα=79019.848 Watts

Also, Mechanical power = Input power-$3I^2R$ =79534.116 Watts **Answer**

Prob.2.2.12 Synchronous motor -maximum torque

A 1492 kW ,unity power factor,3-phase ,star-connected ,2300 V,50 Hz ,30 poles, synchronous motor has a asynchronous reactance of 1.95 ohm/phase. Compute the maximum torque in N-m which this motor can deliver if it is supplied from a constant frequency source and if the field excitation is constant at the value which would result in unity power factor at rated load.

Assume that the motor is of cylindrical rotor type. Neglect all losses.

Solution. Rated kVA,3-phase = 1492

Rated kVA/phase = 497.333

Rated Voltage/phase = V = 1327.906

Rated current, I= 374.524

Synchronous reactance, X_s = 1.950 Ω

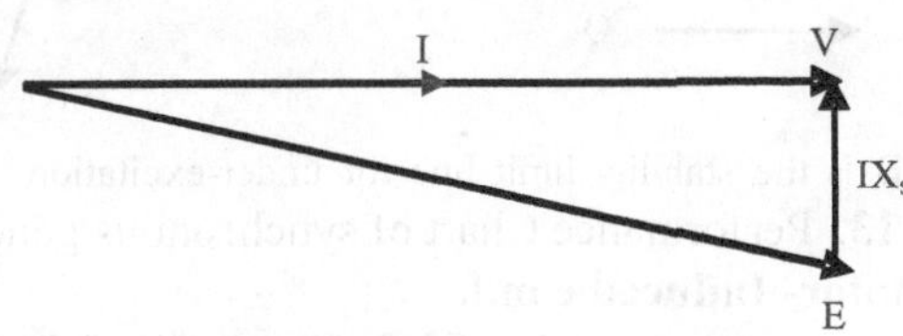

Fig. 2.2.12. *Phasor diagram*

$E = \sqrt{V^2 + I^2 X_s^2}$ = 1515.489 Volts

$P_{max} = EV/X_s$ = 1032.014 kW/phase

P, No. of poles = 30

f, frequency = 50.000 Hz

Speed, N = 120f/P = 200 rpm

Max. Torque, T = $60P_{max}/(2\,\pi\,N)$ = 49.275 kN-m/ph

3-phase max. Torque= 147.825 kN **Answer**

Prob.2.2.13 Synchronous generator -Performance Chart

An alternator is rated at 60 MW, 11 kV, 0.8 pf lagging and has a synchronous reactance of 2 pu and negligible resistance. Choosing a 3-phase MVA scale of 1 cm = 10 MVA, draw the performance chart for this machine and show on it the stability limit for under-excitation allowing a safety margin of 10 %.

Solution: V_{base} =11kV

pf =0.8 lag

MW =60

MVA_b =75

X_{base} = $V_{base} \times V_{base}/MVA_b$ = 1.613 Ω

X_{pu} =2 pu

X = $X_{pu} \times X_{base}$ = 3.226 Ω / phase

V_{ph} = 6.350 kV

I = V_{ph}/X = 1.968 kA

Leading Vars / phase = $V_{ph} \times V_{ph}/X$ = 12.5 MVAr
Leading Vars/3-phase = 37.5 MVAr
AB is the stability limit line for under-excitation

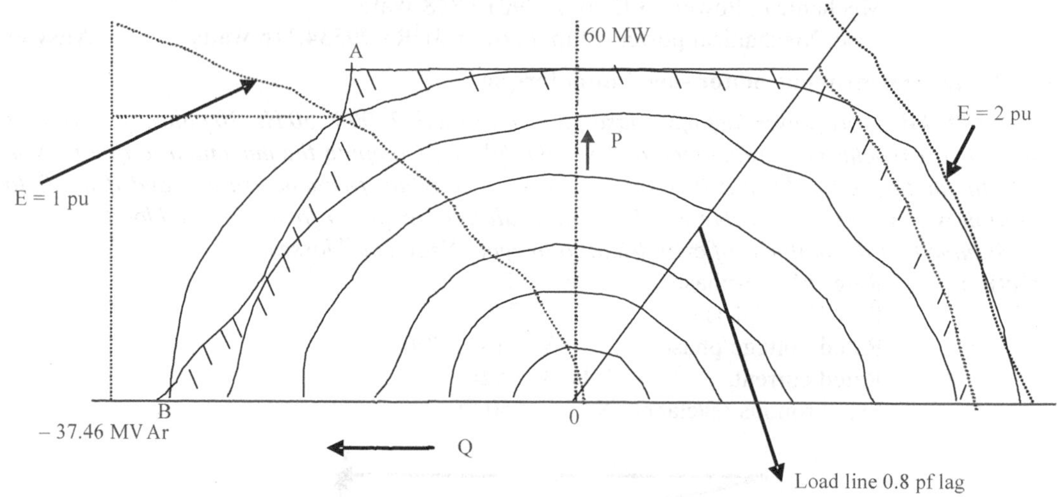

AB is the stability limit line for under-excitation

Fig. 2.2.13. Performance Chart of synchronous generator

Prob.2.2.14 Synchronous motor- Induced e.m.f.

A 3-phase 400 V synchronous motor takes 72.5 A at a p.f. of (i) 0.8 lead, and (ii) 0.8 lag. Calculate the power supplied and the induced e.m.f. for the two cases. The motor impedance / phase is 0.2 + j3 Ω.

Solution.

V per phase 230.94 V
I 72.50 A
Cosϕ = pf = 0.8 lag

$$\text{Power supplied in the two cases} = \frac{1.73 VI \cos\phi}{1000} = 40.184 \text{ kW}$$

(i) Leading pf

R = 0.2 Ω/ph
X = 3 Ω/ph

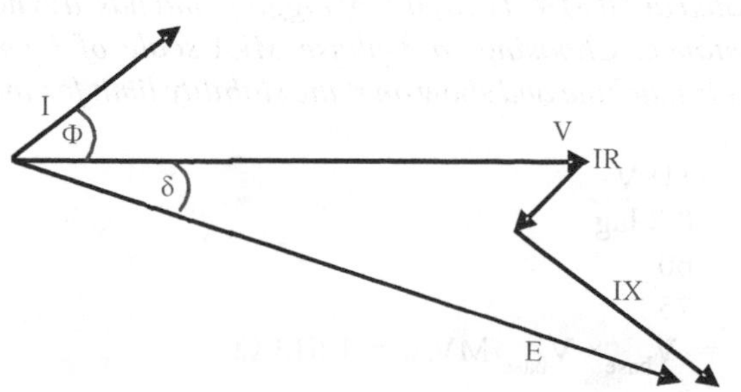

Fig. 2.2.14. (*a*) Phasor diagram for leading power factor

Sinϕ = 0.6
I = I cosϕ +j I sinϕ
Z = R + j X
E = V – IZ

$E = E\cos\delta - j\,E\sin\delta$

$E\cos\delta = \text{V-IR}\cos\phi + \text{IX}\sin\phi = 349.84$

$E\sin\delta = -\text{IX}\cos\phi - \text{IR}\sin\phi = \quad -182.70$

$E = 394.674\ (= \sqrt{E^2\sin^2\delta + E^2\cos^2\delta})$

$E(L-L) = 683.595$ V **Answer**

(ii) Lagging pf

$\sin\phi = 0.6$

$\mathbf{I} = I\cos\phi - jI\sin\phi$

$\mathbf{Z} = R + jX$

$\mathbf{E} = \mathbf{V\text{-}IZ}$

$\mathbf{E} = E\cos\delta - jE\sin\delta$

$E\cos\delta = \text{V-IR}\cos\phi - \text{IX}\sin\phi = 88.84$

$E\sin\delta = -\text{IX}\cos\phi + \text{IR}\sin\phi = -165.30$

$E = 187.661$

$E(L-L) = 325.038$ V **Answer**

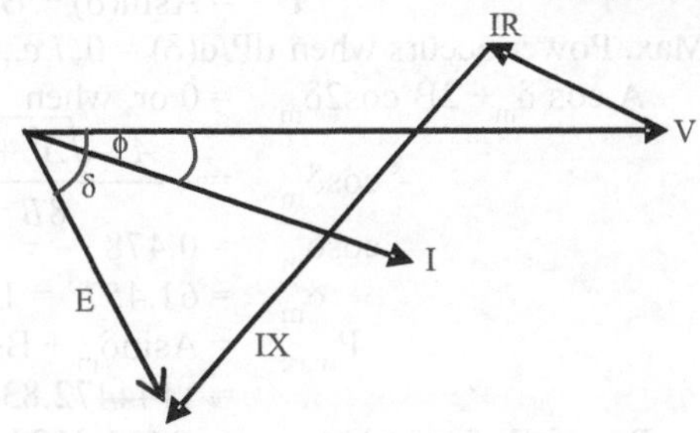

Fig.2.2.14. (*b*) *Phasor diagram for lagging power factor*

Prob.2.2.15 Synchronous motor-Salient-pole-Maximum power

A 1492 kW, unity power factor, 3-phase, star-connected 2200V synchronous motor has reactance of $X_d = 2\ \Omega$ and $X_q = 1\ \Omega$/phase. Neglecting losses, compute the maximum mechanical power in kW which this motor can deliver if it is supplied with electric power from an infinite bus at rated voltage and frequency and if its field excitation is constant that results in unity power factor at rated load.

Solution. Voltage, L – L 2200 V

V, per phase = 1270.171 V

kVA, per phase = 497.333 kVA

I, rated = (kVA/V) × 1000 = 391.548A

Phasor diagram for full-load conditions:

$X_d = 2\ \Omega$

$X_q = 1\ \Omega$

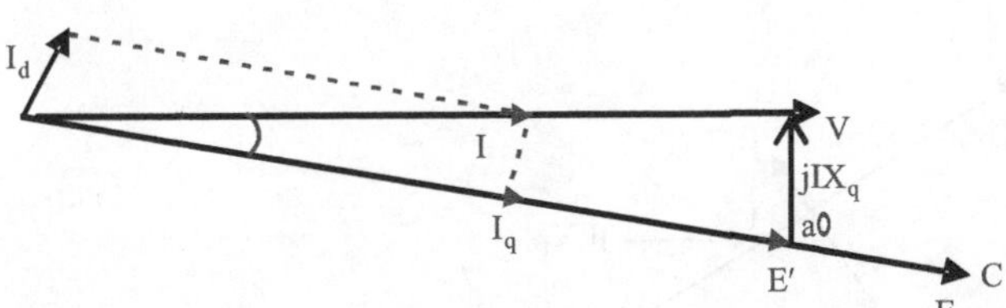

Fig.2.2.15. *Phasor diagram*

$E' = V - jI\,.X_q$

$V = \text{Re}\,E' = 1270.171$ Volts

$\text{Im}E' = -I \times X_q = -391.548$ Volts

$\sqrt{V^2 + IX_q^2}$ E' mag = 1329.152 Volts

$\delta = \tan^{-1}(\text{Im}E'/\text{Re}\,E') = -0.299$ radians $= -17.133°$.

$\sin\delta = 0.295$

$I_d = I \times \sin\delta = 115.344$ A

a'c length $= I_d \times (X_d - X_q)$

a'c = 115.344

$E_{af} = E'$ mag + a'c

E_{af} = 1444.496 Volts

$P = (E_{af} \times V \times \sin\delta/X_d) + V^2(X_d - X_q)\sin 2\delta/2X_dX_q$

$A = E_{af} \times V/X_d$

$A = 917378.728 \quad B = V^2(X_d - X_q)/(2 \times X_d \times X_q)$

$B = 403333.709$

$P = A\sin(\delta) + B\sin 2(\delta)$

Max. Power occurs when $dP/d(\delta) = 0$, *i.e.*, when

$A\cos\delta_m + 2B\cos 2\delta_m = 0$ or, when

$$\cos\delta_m = \frac{-A+\sqrt{A^2+32B^2}}{8B}$$

$\cos\delta_m = 0.478$

$\delta_m = 61.457° = 1.073$ rad

$P_{max} = A\sin\delta_m + B\sin 2\delta_m$

$= 1144472.834$ W

P_{max}, kW for 3 phases = 3433.419 kW **Answer**

Prob.2.2.16 Synchronous machine-Operation at maximum power

A generator having a synchronous reactance of 0.9 pu based on its own rating is connected via a transmission line to a remote bus. The voltage of the remote bus is 1 pu. The line reactance is j0.15 pu per phase based on the machine MVA. The internal e.m.f. E of the machine is kept at 1.35 pu

(a) Calculate the pull-out power of the machine when operated as a generator

(b) At the moment of pull-out, calculate the reactive powers at the generator terminals and at the remote bus.

(c) What is the terminal voltage of the generator at the moment of pull-out ?

(d) Is the machine current overloaded at pull-out ?

Solution.

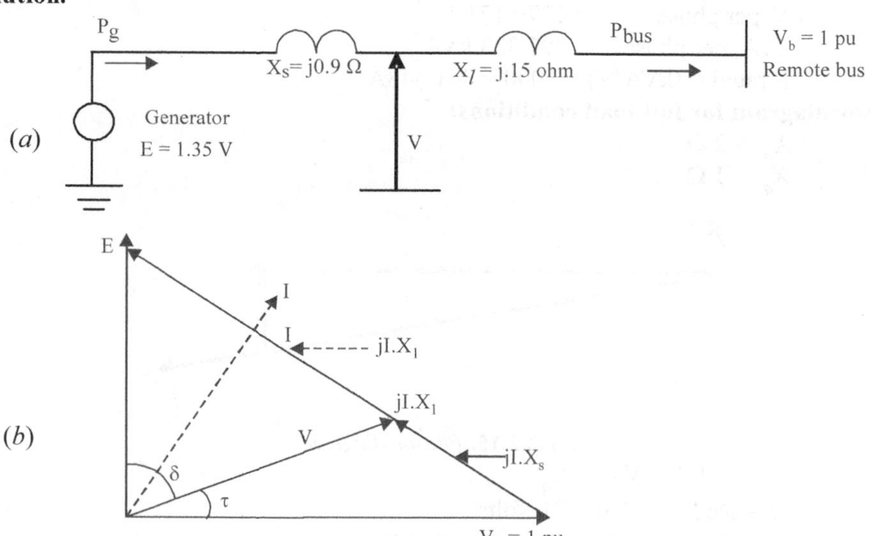

Fig.2.2.16. *(a) Circuit diagram (b) Phasor diagram*

$P_g = P_{bus} = EV\sin\delta/X_s = EV\sin(90-\tau)/X_s = EV\cos\tau/X_s = V V_b \sin\tau/X_l$.

Therefore,

E = 1.35 Volts

X_l = 0.15 Ω

V_b = 1 volts

X_s = 0.9 Ω

$\tan \tau = EXl/V_b.X_s = 0.225$

$\tau = 0.221$ radian

$\delta = (3.14/2) - \tau = 1.57 - \tau$

$\delta = 1.349$ radian

$$I = \frac{\sqrt{E^2 + V_b^2}}{X_s + X_l} = 1.6\ \text{A (p.u)}$$ **Answer**

$$IX_s^2 = E^2 - V^2 - (2EV\cos(1.57 - \tau))$$

This is a quadratic equation in V

$aV \times V + bV + c = 0$

where $a = 1.000$

$b = -2 \times E\cos(1.57\text{-}\tau)$

$b = -0.595$

$c = E^2 - I^2 \times X_s^2$

$c = -0.251$

$$V = \frac{-b + \sqrt{b^2 - 4ac}}{2a}$$

$= 0.880$ Volts (p.u) **Answer**

(*a*) Pull-out power $= \dfrac{EV_b}{X_s + X_1} = 1.286$ **Answer**

(*b*) $$Q_g = \frac{EV\cos\delta}{X_s} - \frac{V^2}{X_s}$$

$= -0.570$ pu **Answer**

Generator behaves like shunt reactor

Reactive power at bus (defined positive from bus),

$Q_b = V_b^2/Xl - (V.V_b \text{Cos}(\tau)/X_l$

$= 0.942$ pu **Answer**

$V = 0.88$ pu **Answer**

$I = 1.6$ pu **Answer**

Prob.2.2.17. *A 500 kVA, 3300 V, 3-phase alternator has a synchronous reactance of 0.25 pu. It is delivering full load at 0.8 pf lagging. By changing the excitation, the e.m.f. is increased by 20 % at this load. Calculate the new current and the power factor. The alternator is connected to infinite bus bars at this load. Calculate the new current and the power factor.*

Solution.

kVA per phase	= 166.667
Voltage per phase	= 1905.257
V, Terminal Voltage	= 1.000 pu
X_s, syn.reactance	= 0.25 pu
I, full load current	= 1pu

E, Excitation voltage $= V + jI \times X_s$

$\cos\phi = \text{pf} = 0.8$ lag

$\sin\Phi = -0.6$

$\text{Re}\,E = V - IX_s \sin\phi = 1.15$

$\text{Im}\,E = I \times X_s \cos\phi = 0.2$

$E\,\text{mag} = 1.167$ pu $(= \sqrt{(\text{Re}\,E)^2 + (\text{Im}\,E)^2})$

On changing the excitation, the new labels are E', I', $\cos\phi'$, $\sin\phi'$, δ'. E' leads V by δ'

$$E' = 1.2 \times E = 1.401 \text{pu}$$
$$E' = V + jI'X_s$$
$$\text{Re } \mathbf{E}' = V - I' \times X_s \times \sin\phi'$$
$$\text{Im } \mathbf{E}' = I' \times Xs \times \cos\phi'$$
$$I' \cos\phi' = \text{Im}E'/X_s = 1.4 \times \sin\delta'/X_s$$
$$= 5.6 \sin\delta' \text{ since } X_s = 0.25$$
$$I' \sin\phi' = (V\text{-Re } E')/X_s$$

Since power remains the same'

$$I' \cos\phi' = I \times \cos\phi$$
$$0.8 = 5.6 \times \sin\delta'$$
$$\delta' = 0.143 \text{ radian}$$
$$I' \sin\phi' = 4\text{-}5.6 \times \cos\delta' = -1.543$$
$$I' = \sqrt{I'^2 \cos'^2 \phi + I'^2 \sin^2 \phi'} = 1.738 \text{ pu}$$ **Answer**

$\cos\phi'$, new *pf* = 0.460 lag **Answer**

I full load, Amps = kVA/Voltage = 87.477 A

I', Amps = 152.007 A **Answer**

Prob.2.2.18. *A 3-phase generator supplying 2800 kW at power factor 0.7 lagging is loaded to its full capacity, i.e. its maximum rating in kVA. If the power factor is raised to unity by means of an overexcited 3-phase synchronous motor, how much more active power can the generator supply and what must be the power factor of the synchronous motor, assuming that the latter absorbs all extra power obtainable from the generator? Sketch the phasor diagram.*

Solution.

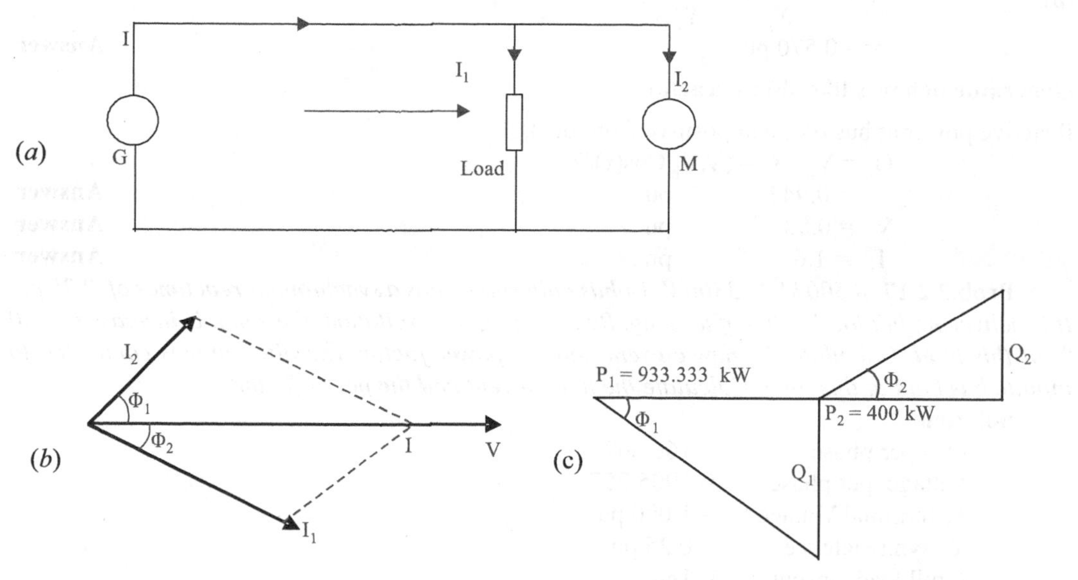

Fig.2.2.18. (*a*) *Circuit diagram*,(*b*) *and* (*c*) *phasor diagrams*

KW per phase = 933.333

$\cos\phi_1$ = pf = 0.7 lag

kVA / phase = kW/cos ϕ_1 = 1333.333

Full load / phase = kVA / phase = 1333.333kW

Generator can supply an additional power, P_2 = 1333.333 – 933.333 = 400kW/phase

= 1200 kW (3 –phase) **Answer**

$\tan\phi_1 \quad = 1.020$

$Q_1 \quad = P_1\tan\phi_1 = 952.190$

$Q_2 \quad = P_2\tan\phi_2 = 400\tan\phi_2$

Since $Q_1 = Q_2$,

$\tan\phi_2 \quad = 2.380$

$\phi_2 \quad = 1.173$ radian

$\cos\phi_2 \quad = 0.387$ lead **Answer**

Prob.2.2.19 Synchronous generator-Regulation from Magnetization curve

The magnetization curve of a 400 V, 50 Hz alternator is given by the following readings:

Open circuit, Volts:	*Field current*: *A*
266	2.0
334	2.5
377	3.0
422	3.5
450	4.0
484	4.5
508	5.0

Full - load current is obtained by 2A excitation on short-circuit. Calculate the percentage regulation at 0.8 pf lagging.

Solution:

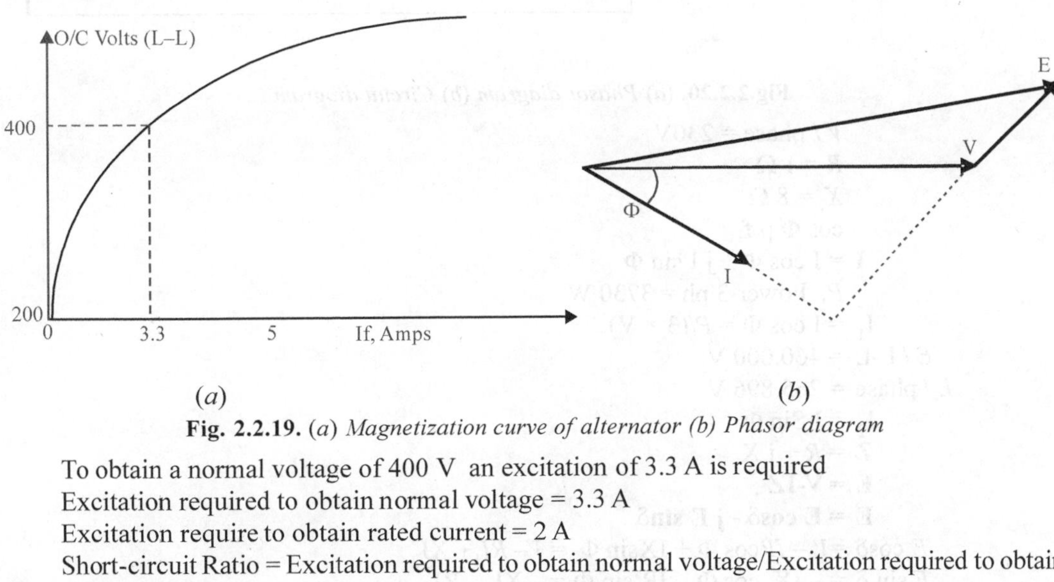

(*a*) (*b*)

Fig. 2.2.19. (*a*) *Magnetization curve of alternator (b) Phasor diagram*

To obtain a normal voltage of 400 V an excitation of 3.3 A is required

Excitation required to obtain normal voltage = 3.3 A

Excitation require to obtain rated current = 2 A

Short-circuit Ratio = Excitation required to obtain normal voltage/Excitation required to obtain rated current = 1.65

X_s, Syn. Reactance = 1/short-circuit Ratio

$X_s \quad = 0.606$ pu

$V_{\text{base}} \quad = 400/1.73205$

$= 230.940$ Volts

$V \quad = 1$ pu

$I \quad = 1$ pu

$X_s \quad = 0.606$ pu

$\cos\phi \quad = 0.8$

$\sin\phi \quad = 0.6$

$$E = \sqrt{(V\cos\phi)^2 + (V\sin\phi + IX_s)^2}$$

E_{pu} = 1.447

E, Volt = $E_{pu}V_{base}$= 334.220

Regulation =

= (E–V)/V = 0.447

= 44.72% **Answer**

Prob.2.2.20 Synchronous motor-Current & Power factor

A 400 V, 3-phase star-connected synchronous motor takes 3.73 kW at normal voltage , and has an impedance of $1+j8\ \Omega/phase$. Calculate the current and the p.f. if the induced emf is 460 V.

Solution: In this problem, E is greater than V. Therefore, p.f. is leading.

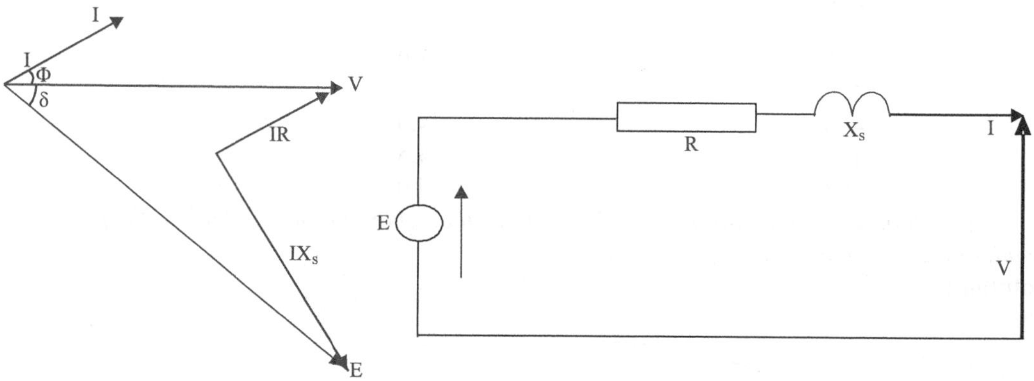

Fig.2.2.20. (*a*) *Phasor diagram* (*b*) *Circuit diagram*

V/ phase = 230V

$R = 1\ \Omega$

$X_s = 8\ \Omega$

cos Φ p.f.

I = I cos Φ + j I sin Φ

P, Power-3 ph = 3730 W

I_1 = I cos Φ = $P/(3 \times V)$

E / L-L = 460.000 V

E / phase = 265.896 V

I_2 = I Sin Φ

Z = R + j X

E = V-IZ

E = E cosδ - j E sinδ

$E\cos\delta = V - IR\cos\Phi + IX\sin\Phi = V - RI_1 + XI_2$

$E\sin\delta = -IX\cos\Phi - IR\sin\Phi = -XI_1 - RI_2$

$$E^2 = (V - I_1R + I_2X)^2 + (X_sI_1 + RI_3)^2 \qquad \dots (1)$$

$$P = 3 \times V \times I_1 \qquad \dots (2)$$

Eliminate I_1 from Eqs. (1) and (2) respectively, and obtain

$$E^2 = \left(V - \frac{RP}{3V} + X_sI_2\right)^2 + \left(\frac{X_sP}{3V} + RI_2\right)^2$$

This is a quadratic equation in I_2

$aI_2^2 + bI_2 + c = 0$,

where $a = R^2 + X_s^2 = 65$

$b = 2 \times VX_s = 3680$

$$c=V^2-E^2-\frac{2}{3}RP+\frac{R^2P^2}{9V^2}+\frac{Xs^2P^2}{9V^2}=-18387.853$$

$$I_2=\frac{-b+\sqrt{b^2-4ac}}{2a}$$

Therefore,

$I_2 = I\sin\phi = 4.620 = I\sin\phi$

$I_1 = I\cos\phi = 5.410 = I\cos\phi$

$I = \sqrt{I_1^2+I_2^2} = 7.114$ **Answer**

$\cos\Phi = I_1/I = 0.760$ lead **Answer**

Prob. 2.2.21. Synchronous generator-Salient pole-Performance

A 6.6 kV, 5 MVA, 6-pole 50, Hz star-connected generator has X_d= 8.7 Ω / phase, X_q =4.35 Ω / phase. If the excitation is so adjusted that the excitation voltage E_f= 11 kV(L-L), and the load angle is 30° (elec.), determine

(a) the power factor, output current, and power in per unit

(b) the load angle at maximum torque

(c) the ratio between maximum torque and that occurring with δ = 30°.

(d) the stiffness in Nm per mechanical radian for a load angle of 30°.

Solution. ***(a)***

kVA / phase = 1666.666

V / phase = 3815.028

I_{fl} = kVA × 1000/V = 436.8686869 A

load angle δ = 30° = 0.523 radian

sinδ = 0.499

X_d = 8.7 Ω

X_q = 4.35 Ω

E_f per ph = 635.838 Volts

$$I_q = \frac{V\sin\delta}{X_q}$$

= 438.307A

$I_d = (E_f - V\times\cos\delta)/X_d$

= –306.733

$I = \sqrt{I_d^2\times I_q^2}$

= 534.975 A

Ipu = I/I_{fl} = 1.224pu **Answer**

$I\cos\Phi = I_q\cos\delta + |I_d|\sin\delta$

$$\cos\Phi = \frac{I_q\cos\delta + I_d(Mag.)\sin\delta}{I}$$

= 0.996 lag **Answer**

P, Power = Vpu × Ipu × cosΦ

= 1.219 pu **Answer**

(b)

$$T=\frac{VE_f.\sin\delta}{X_d}+\frac{V^2}{2}\left(\frac{1}{X_q}-\frac{1}{X_d}\right)\sin 2\delta$$

Differentiating this expression with respect to δ and putting it equal to zero gives

$$\cos\delta = \frac{E_f}{2V} X_q \frac{1}{X_d - X_q} + \sqrt{\frac{E_f}{2\times 4V}\cdot\frac{E_f}{2V}\times\left(\frac{X_q}{X_d - X_q}\right)^2 - \frac{1}{2}}$$

$$\frac{E_f}{2V} = 0.0833$$

$$\frac{X_q}{X_d - X_q} = 1$$

Therefore ,

$\cos\delta$ = 0.458

δ (max. value) = 62.752° **Answer**

= 1.095 radians

δ(max) = 1.095

$\sin\delta$ = 0.889

$\sin 2\delta$ = 0.814

(c)

$$\text{Tmax} = \frac{VE_f \sin\delta_{max}}{X_d} + \frac{V^2}{2}.\sin 2\delta_{max}\left(\frac{X_d - X_q}{X_d.X_q}\right) = 928818.912$$

T, rated, at 30° = 863521.838

T_{max}/T,rated = 1.075 **Answer**

f, frequency = 50

P, pole pairs = 3

N_s, = f/p =16.666 rev/s

$w_s = 2\pi n_s$ = 104.666 rad/sec

(d)

$$\frac{dT}{d\delta} = \left[\frac{VE_f \cos\delta}{X_d} + \frac{V^2(X_d - X_q)}{X_d.X_q}\cos 2\delta\right]\frac{1}{\omega_s}$$

at 30° dT/d δ =10306.378 / phase

for 3-phases, dT/d δ, = 30919.135 Nm per electrical radian

Nm/mechanical radian = 30919.135 × p = 92757.407 **Answer**

Prob.2.2.22

Synchronous generator -Parallel operation

Two identical 16.66MVA, 11 kV 3 phase alternators are in parallel equally excited and sharing equally a load of 25 MW at 11 kV and 0.8 pf lagging. Calculate the excitation at rated voltage of each alternator assuming a constant synchronous reactance per phase of 1.08 pu. If the excitation voltage of one alternator is to be reduced by 15 % determine by how much the excitation voltage of the other alternator must be increased, in order to avoid a change of terminal voltage and of water supply to either set.

Solution.

(a)

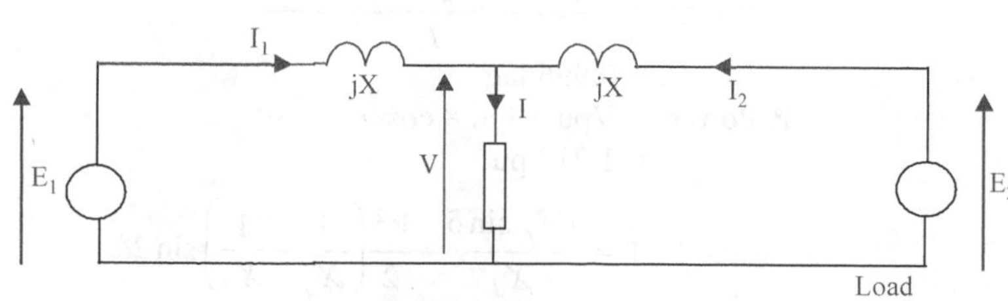

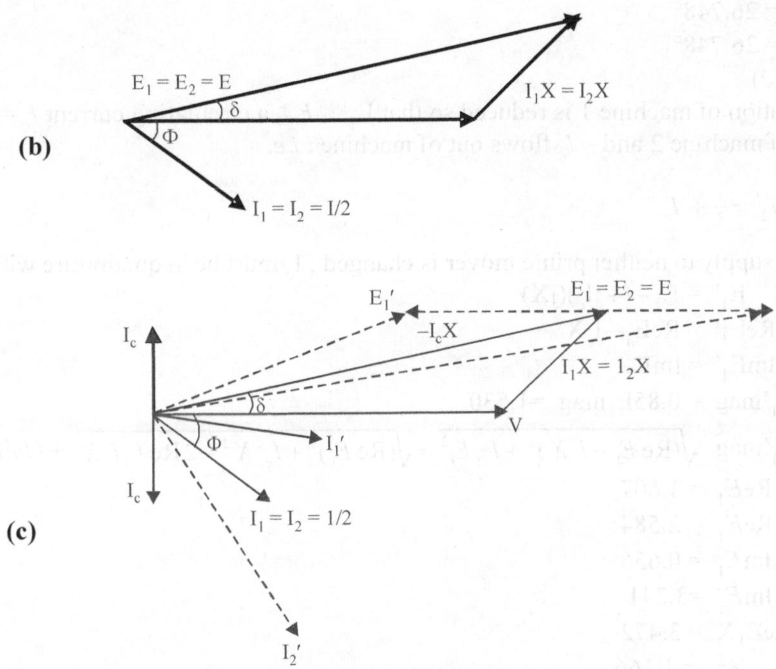

Fig.2.2.22 *(a) Circuit diagram (b) Phasor diagram (c)Phasor diagram when excitation is reduced*

$\mathbf{I_1} = (\mathbf{E_1} - \mathbf{V})/jX \quad \mathbf{I_2} = (\mathbf{E_2} - \mathbf{V})/jX, \quad I = I_1 + I_2$

$\mathbf{I} = (\mathbf{E_1}/jX) + (\mathbf{E_2}/jX) - (2\mathbf{V}/jX)$

$\mathbf{V} = (\mathbf{E_1} + \mathbf{E_2} - \mathbf{I}jX)/2, \ (\mathbf{V}/jX) = (\mathbf{E_1}/2jX) + (\mathbf{E_2}/2jX) - (\mathbf{I}/2)$

$\mathbf{I_1} = \mathbf{E_1}((1/jX) - (1/2jX)) - (\mathbf{E_2}/j2X) + (I/2) = (\mathbf{E_1} - \mathbf{E_2})/2jX) + \mathbf{I}/2$

$\mathbf{I_2} = \mathbf{E_2}((1/jX) - (1/2jX)) - (\mathbf{E_2}/j2X) + (I/2) = (\mathbf{E_2} - \mathbf{E_1})/2jX) + \mathbf{I}/2$

Here, $\mathbf{E_1} = \mathbf{E_2}$

$\mathbf{I_1} = \mathbf{I_2} = \mathbf{I}/2$

P Load = 25 MW

V L–L = 11kV

$\cos\phi = pf = 0.8$ lag

$$I = 1642.143 \text{ A} \left(\frac{P}{V \times pf}\right)$$

$I_1 = 821.071\text{A}$

$I_2 = 821.071\text{A}$

MVA_b base = 16.66 MVA

V_b base = 11kV

I_b base = 875.459 A

$I_1pu = 0.937$

$I_2pu = 0.937$

Vpu = 1

$\mathbf{E_1} = V + j\mathbf{I1}X$

X = 1.08pu

Re $\mathbf{E_1} = V + I_1 \sin \Phi X = 1.6077$

Im$\mathbf{E_1} = I_1 \cos\Phi . X = 0.810$

$\mathbf{E_1}$mag = 1.800 pu

E_1 angle $= 26.748°$

$\delta = 26.748°$

See Fig.2.2.22(*c*)

When the excitation of machine 1 is reduced so that $E_1' < E_2'$, a circulating current $I_c = (E_2' - E_1') / j2X$ flows out of machine 2 and $-I_c$ flows out of machine , *i.e.*

$$I_1' = \frac{I}{2} - I_c \qquad I_2' = \tfrac{1}{2} + I_c$$

Since the water supply to neither prime mover is changed , I_c must be in quadrature with V.

$E_1' = E_1 - (-jI_c)(jX)$

$\text{Re}E_1' = \text{Re}E_1 - I_cX$

$\text{Im}E_1' = \text{Im}E_1$

E_1'mag $= 0.85E_1\text{mag} = 1.530$

E_1'mag $\sqrt{(\text{Re}\,E_1 - I_cX)^2 + I_M E_1^2} = \sqrt{(\text{Re}\,E_1)^2 + I_c^2X^2 - 2\,\text{Re}\,E_1 I_c X} + (ImE_1)^2$

$\text{Re}E_1 = 1.607$

$\text{Re}E_1\text{Re}E_1 = 2.584$

$\text{Im}E_1\text{Im}E_1 = 0.656$

$\text{Re}E_1\text{Re}E_1 + \text{Im}E_1\text{Im}E_1 = 3.241$

$2\text{Re}E_1X = 3.472$

$X^2 = 1.166$

Therefore, $1.1664I_c^2 + 3.241 - I_c3.472725 = 2.341$ ($= (E'_1\,\text{mag})^2$)

Quadratic, $aI_{c2} + bI_c + c = 0$ where

$a = 1.166$

$b = 3.241$

$c = 0.899 = (3.241 - 2.341)$

Therefore, $I_c = 2.466$

$\mathbf{E_2}' = \mathbf{E_1} + 2\mathbf{I_c}jX = \text{Re}E_1 + j\,\text{Im}E_1 + 2jI_cjX$

$\text{Re}E_2' = \text{Re}E_1 - 2IcX = 3.719$

$\text{Im}E_2' = \text{Im}E_1 = 0.810$

E_2'mag 3.806 pu **Answer**

CHAPTER 3

INDUCTION MOTOR

3.1 INTRODUCTION

For induction motor simulation see the websites

- http://eagle.ca/~cherry
  (Cherry Tree Scientific Software – by Graham J. Rogers)
- Menu-Driven Graphical User Interface (GUI) Programs
  – By Prof. H. Sadat can be seen at http://www.msoe.edu/~saadat/matlabgui.htm.

The **transformer program** obtains the transformer equivalent circuit from open-circuit and short-circuit tests. It also finds the transformer performance characteristics using the transformer parameters. In the **induction motor program the** parameters of the equivalent circuit can be entered on the circuit diagram. The program draws the Thevenin's equivalent circuit. Performance characteristics are then obtained for the specified slip and the torque slip curve is plotted on the screen. The analysis can be repeated for different slips. A GUI program deals with the design of a **first-order controller** in the forward path of a closed-loop control system. This GUI program designs the following controllers: Proportional, phase-lag, phase-lead, and PD, PI, and PID controllers. To run the controller design program at the MATLAB prompt type >> rldesigngui.

3.1.1 Basic Theory & Construction

Introduction to the concept of rotating magnetic field and production of rotating magnetic flux by two-phase and three-phase currents flowing through concentrated windings (1slot / pole / phase), magnitude & speed of resultant rotating flux, constructional features of induction motors, three-phase windings

Expression for the stator induced e.m.f., principle behind the development of torque by the rotor, rotor frequency & slip.

3.1.2 Phasor Diagram & Equivalent Circuit

Expressions for rotor current, establish the relationship-rotor input: copper loss: output-1: s: (1-s)

3.1.3 Steady-State Analysis

When a voltage is applied to the stator of a 3-phase induction motor causing 3-phase currents to flow, a rotating m.m.f. and consequent flux is set up in the air gap. This flux moves at synchronous speed (passes one double-pole pitch in one period), cuts the short-circuited rotor conductors and sets up in them an e.m.f.; and consequently a current. This current interacts with the flux to produce a torque which moves the rotor in the same direction as the field motion. As the motor speeds up , the relative motion between the field and the conductors falls, becoming zero at synchronous field, so that the rotor can never quite attain this (although it nearly does so under normal load conditions).

By considering the above explanation, students can visualise the general effect of speed (or slips) on the rotor e.m.f.(sE2),rotor frequency (sf1),rotor leakage reactance(sx2),power factor and torque (proportional to rotor current, power factor and flux).

The stator has resistance and leakage reactance which cause voltage drops, so that, with a constant applied voltage V, the e.m.f. E_1, and therefore the flux, falls as the current (load) increases. Also the core requires magnetization, so that the stator carries a magnetizing current Im which has no counterpart in the rotor. The stator core has iron losses (hysteresis and eddy current) requiring a further current Ic. The core loss in the rotor is negligible except at low speeds.

3.1.4 Prediction of behaviour

The behaviour of the machine can be fully determined :

Mechanical output= (1-s) (rotor input)

Torque (Nm) = Mechanical Output /ωr = Rotor Input /ω

Rotor loss (I^2r) = s × (rotor input)

Since the rotor input is $(I_2')^2 r_2'/s$, output, losses and torque can readily be calculated for any value of slip.

It is convenient to split the resistance r_2'/s in the equivalent circuit into two components r_2' and $r_2'(1-s)/s$ the power dissipated in which represents the rotor loss and the mechanical output respectively.

Although induction machines normally operate as motors with *s* between 0 (synchronous speed) and 1 (standstill), conditions with *s* between 1 and 2 (braking by plugging) and s negative (generating) must also be considered. The procedure is illustrated in Example 3.2.16, below.

3.1.5. Torque Speed Characteristics

Effect of rotor resistance variation, maximum torque and the slip at which it occurs

3.1.6 Testing & Circle Diagram

No load and blocked rotor tests equivalent circuit parameters, circle diagram.

Students should be aware of the simple no-load and short-circuit (locked-rotor) tests, since results from these are commonly used for determining the constants for use in the equivalent circuit or for drawing the circle diagram. The no-load test ($I_2'=0$) enables r_μ and x_μ to be found; with the locked-rotor test, usually taken at a low voltage, the no-load current Io can be neglected, so that (r_1+r_2') and (x_1+x_2') can be found; r_1 can be measured separately enabling r_2' to be determined. Note that the power on the no-load test includes the friction loss, whereas this should theoretically be regarded as mechanical output. The circle diagram can very easily be drawn from the above test results.

3.1.7 Starting, Braking, Speed Control and Generation

Requirements to develop high starting Torque and reduced starting current, Deep bar and double -cage rotors, Starters (star-delta, auto transformer & rotor resistance starters)

Performance under all the above conditions can readily be determined from the equivalent circuit.

Starting devices include added rotor resistance, star/delta switching, autotransformers and high-torque rotors; braking schemes include regenerative braking (driving at above synchronous speed), plugging (interchanging two of the supply leads) and d.c. injection (disconnecting the stator winding and injecting into it a direct current).

Calculations on speed control by added rotor resistance can be made directly from the equivalent circuit. The principle of slip –energy –recovery control schemes can be understood using the following power relations:

Rotor power input:	Slip power(rotor I^2r loss):	Mechanical power
=1	: s	: (1-s)
= ω	: $(\omega - \omega_r)$	: ω_r

where ω is the synchronous speed and ω_r the motor speed (rad/s).

Many students become confused in considering the power factor of the induction machine when running as a generator. As a motor the machine draws both kW and lagging kVAr from the supply, the latter being necessary to magnetise the machine; the power factor therefore is lagging. As a generator the machine still has to draw approximately the same kVAr from the supply, but the kW flow has been reversed and is delivered to the supply; the power factor is therefore leading. The conditions are thus similar to those obtaining with an underexcited synchronous generator which also has to draw some of its magnetisation from the supply, since that provided by the field excitation is insufficient.

3.1.8 Construction

Students should be aware of the main constructional features of slip-ring and cage machines and should understand qualitatively some of the 'second order 'problems that arise. These include the main design features such as the need for a very small air gap, nature of leakage-flux paths and the importance of leakage reactance, significance of m.m.f. harmonics, slotting and its effect on noise, unbalanced magnetic pull and self excitation by capacitance.

3.1.9 Other Modes of Operation

Braking of induction motor, induction generator

3.1.10 Single-phase Induction Motor

Double revolving field theory, Torque-speed characteristics, Equivalent circuit, Starting methods, Auxiliary winding parameters for maximum starting torque, Characteristics of split-phase motor, capacitor start motor, two value capacitor motor, shaded pole motor

The 1-phase motor can be treated by the cross-field or the rotating field theories, the latter being the more usual.

The rotating field theory is based on the fact that a pulsating m.m.f. can be resolved into two rotating m.m.f.s moving at synchronous speed in opposite directions; the motor can thus be treated as a combination of two 3-phase motors, one producing a forward torque as in a normal motor, and the other a backward torque as in a motor operating between $s = 1$ and $s = 2$. Development of this idea leads to a torque /slip curve of similar shape to that of a 3-phase motor, except that the torque is zero at $s = 1$(no starting torque), and also indicates other minor differences in characteristics. It also leads to the equivalent circuit shown in Prob.3.2.12 given in the problems section.

3.1.11 Two-phase Servomotor

The servomotor used as an actuator in a closed-loop control system must have good controllability, linearity, sensitivity and a high torque/inertia ratio. The 2-phase induction motor fulfills these requirements with reasonable accuracy, although its output is limited to about 0.5 kW.

One winding (reference winding) of the motor is supplied at a constant voltage V_1 and the other with the variable control signal displaced in time phase by 90°. If the voltage on the control winding is equal to, or equivalent to, V_1 a normal induction motor characteristic might be expected (curve A of Fig.3.2.17 a in Prob.2.1.17). This is widely nonlinear but, by designing the motor with a high ratio of r/x for the rotor, curve B can be obtained which has reasonable linearity. At various values of control voltage V_2 it can be assumed that the linear series of curves in Fig.3.2.17 (b) can be obtained.

While the behaviour of the motor can be studied by normal methods, this is not usually needed as, with control systems, it is the transfer function that is generally required, *i.e.*, the relation between motor shaft displacement θ and the control voltage V_2.

3.1.12 Induction Motor torque-speed Characteristics -Demonstration Module

This module demonstrates the torque versus speed characteristics of a three-phase induction motor. Draw the exact equivalent circuit of the motor and observe the effects of changing various parameters on the torque versus speed curve. In particular, vary the rotor resistance, applied voltage, and the supply frequency and observe the effects on the torque-sped curves. Also verify whether the motor can supply a particular load. Draw two types of plots: torque versus slip and torque versus speed. The type of plot can be chosen by selecting the desired plot type from the plot pull-down menu under commands. There is a list of parameters of the induction motor that can be changed by clicking on the values and editing them to the desired values. The curve is updated by clicking on the calculate button. The curve is drawn for the displayed set of parameters.

This calculation module demonstrates the torque versus speed characteristics of a three -phase induction motor. The student is shown the exact equivalent circuit of this motor. The student can observe the effects of changing various parameters on the torque versus speed curve. In particular, the student is asked to vary rotor resistance, applied voltage, and the supply frequency, to observe the effects on the torque versus speed curves. The student can also verify whether the motor can supply a particular load.

The student draws two types of plots: torque vs. slip or torque vs. speed .The type of plot can be chosen by selecting the desired plot type from the plot pull-down menu under commands .There is a list of parameters that pertain to the induction motor. Any of the parameters can be changed by clicking on the values and editing them to the desired value. The curve is updated by clicking on the calculate button. The curve is drawn for the displayed set of parameters. The *hold-on* and *hold- off* buttons are useful in fixing whether the previous curves are erased or preserved on the display. By clicking on *hold-on, the* previous plots can be preserved, while the new plots are superimposed. This makes it possible to compare two curves. The *get co-ords* button is useful in getting the coordinates of any points on the plots. Under parameter list, the last parameter is the load type. A quadratic type of load is assumed .The three fields under the load correspond to the three co-ordinates in a load of type: $T_{load} = an^2 + bn + c$, where n is speed in rpm.

The student is asked the following sample questions:

* The load on the motor is given by $T_{load} = 10n + 5000$. Can we operate the motor with this load ?

If so, find the speed at which the motor operates. Find also if the motor can be started.

- Find the external rotor resistance needed to get the maximum starting torque.

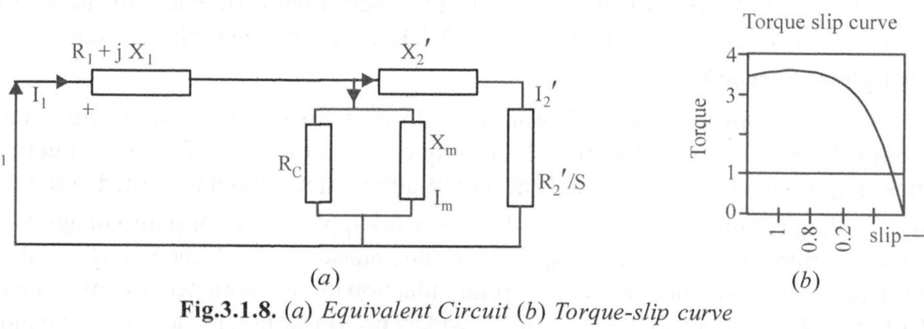

Fig.3.1.8. *(a) Equivalent Circuit (b) Torque-slip curve*

Data:

R_1=1 Ω, R_2'=4 Ω ,X_1=X_2'= 4 Ω

R_c=200 Ω, X_m= X_{phi}= 50 Ω

Voltage (V) =120 Volts

ω_s (rad/s) =188.5

Calculate :

$$V_{1a} = \frac{V_1 . jX_p \phi}{R_1 + jX_1 + jX_\phi} \quad \text{Re} V_{1a} = V_1 X_1 X_\phi \frac{X_\phi{}^2}{R_1{}^2 + (X_1 + X_\phi)^2} = 64.128$$

$\text{Im} V_{1a} = 1.187$

$$V_{1a}(Mag) = \sqrt{(\text{Re} V_{1a})^2 + (I_m V_{1a})^2} = 64.139$$

$$R_{e1} + jX_{e1} = \frac{(R_1 + jX_1) jX_{\phi 2} X}{R_1 + jX_1 + jX_\phi}$$

$$R_{e1} = \frac{X_1 X_\phi (X_1 + X_\phi) + R_1{}^2}{(R_1{}^2 + (X_1 + X_\phi)^2} = 3.702$$

$X_{e1} = -0.0514$

Slip = =1 = 0.8 = 0.6 = 0.4 = 0.2 = 0.0001

$$I_2 = \frac{V_1 a(Mag)}{\sqrt{(\text{Re}_1 + R_2 / s)^2 + (X_2 + X_e)^2}}$$

=7.409 = 6.711 = 5.780 = 4.497 = 2.669 = 0.0016

$\frac{3I_2{}^2 R_2}{\omega_s . s}$ = 3.495 = 3.584 = 3.545 = 3.219 = 2.267 = 0.0016

speed n = load 0 = 300 = 600 = 900 = 1200 = 1499.85

a = 0 = 0 = 0 = 0 = 0 = 0

b = 0 = 0 = 0 = 0 = 0 = 0

c = 1 = 1 = 1 = 1 = 1 = 1

$T_{load} = an^2 + bn + c$ = 1 = 1 = 1 = 1 = 1 = 1

T_{max} occurs at $s = R_2/X_2 = 1$

External rotor resistance needed to get max. starting torque = $X_2 - R_2 = 0$

OBJECTIVE TYPE QUESTIONS

1. The speed at which the rotating magnetic field produced by stator currents rotates in an induction motor is:
(a) Synchronous speed (b) Rotor speed
(c) Greater than synchronous speed
Ans. (a)

2. A 3-phase 50 Hz induction motor runs at speed of 940 rpm the speed of the rotating magnetic field will be:
(a) 940 rpm (b) 1000 rpm
(c) 1050 rpm (d) 1100 rpm
Ans. (b)

3. The no-load speed of an induction motor is of the order of
(a) 8 % (b) 4 %
(c) 2 % (d) 6 %
Ans. (c)

4. Iron losses in the rotor of a 3-phase induction motor are negligible, because
(a) Frequency of rotor e.m.f. is too low
(b) Flux linking the rotor is of constant magnitude
(c) Flux density in the rotor parts is too low
Ans. (a)

5. Full-load copper losses in a 3-phase 50 Hz 4-pole induction motor running at 1455 rpm are 300 W. The rotor input is

(a) 5 kW (b) 10 kW
(c) 20 kW (d) 50 kW

Ans. (b)

6. The no-load current of a 3-phase induction motor in terms of its full-load current is of the order of

(a) 5 % (b) 10 %
(c) 25 % (d) 50 %

Ans. (c)

7. The power factor of a 3-phase induction motor at no load is approximately

(a) 0.2 (b) 0.7
(c) 0.85 (d) 1

Ans. (a)

8. The maximum torque in a 3-phase induction motor occurs at a slip

(a) R_2/X_2 (b) X_2/R_2
(c) R_2^2/X_2 (d) R_2^2/X_2^2

Ans. (a)

9. For high starting torque, the frequently used 3-phase induction motor is

(a) Squirrel-cage type (b) Slip-ring type
(c) Deep bar squirrel-cage type (d) Double-cage induction motor

Ans. (b)

10. The starting torque in a single-phase induction motor is

(a) High (b) Low
(c) Zero (d) Very low

Ans. (c)

11. The efficiency of a single-phase induction motor as compared to a 3-phase induction motor rating is

(a) Lower (b) Higher
(c) Equal

Ans. (a)

12. The starting torque of a 3-phase induction motor can be increased by increasing

(a) the rotor reactance (b) the rotor resistance
(c) the stator resistance

Ans. (b)

13. The maximum starting torque in the induction motor is developed when the rotor reactance is

(a) half of the rotor resistance (b) twice of the rotor resistance
(c) equal to the rotor resistance

Ans. (c)

14. A change of 5% in the supply voltage of an induction motor will produce a change of approximately

(a) 5% (b) 7%
(c) 10% (d) 25% in the rotor torque.

Ans. (d)

15. The transformation ratio of an induction motor can be found by

(a) blocked rotor test (b) open circuit test
(c) slip test (d) stator resistance test

16. The crawling in the induction motor is caused by

(a) improper design of the machine (b) low supply voltage

(c) high loads (d) harmonics developed in the motor.

Ans. (d)

17. Which one of the following motors is most suitable for signalling devices and timers?

(a) d.c. series motor (b) a.c. series motor

(c) induction motor (d) reluctance motor

Ans. (d)

PROBLEMS WITH SOLUTIONS

Prob.3.2.1 Induction motor-Performance from equivalent circuit

A 3-phase star-connected 220 V (L-L) 7.46 kW, 50 Hz, 6-pole induction motor has the following constants referred to the stator:

$R_1 = 0.294\ \Omega$, $R_2 = 0.144\ \Omega$

$X_1 = 0.503\ \Omega$, $X_2 = 0.209\ \Omega$

$X_m = 13.25\ \Omega$

The total friction, windage, and core losses may be assumed to be constant at 403 W, independent of load. Compute the speed, output torque and power, stator current, power factor, and efficiency when the motor is operated at rated voltage and frequency with slip of 2 %.

Solution. Equivalent Circuit

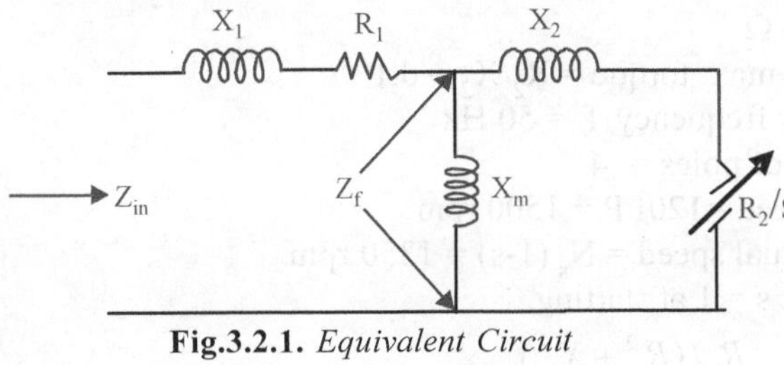

Fig.3.2.1. *Equivalent Circuit*

$R_1 = 0.294\ \Omega$

$R_2 = 0.144\ \Omega$

$X_1 = 0.503\ \Omega$

$X_2 = 0.209\ \Omega$

$X_m = 13.25\ \Omega$

$Z_f = R_f + j X_f = (R_2/s + j X_2)$ in parallel with $j X_m$

Slip, s = 0.02

$R_f = (R_2.X_m^2/s)/((R_2/s)^2 + (X_2 + X_m)^2) = 5.43$ ohms

$$X_f = \left(\frac{R_2.X_2}{s}\right) / \left[\left(\frac{R_2}{s}\right)^2 + \left(X_2 + X_m\right)^2\right] = 3.11\text{ohms}$$

$Z_{in} = R_1 + R_f + jX_1 + jX_f$

Re $Z_{in} = 5.72\ \Omega$

Im $Z_{in} = 3.61\ \Omega$

$Z_{in} = 6.76\ \Omega$

Angle of $Z_{in} = 32.27° = \phi$

Supply voltage , V/phase = 127.02 Volts

I_1, Stator current = V/Z_{in} = 18.78 Amps. **Answer**

Power factor, cos Φ = 0.85 lag **Answer**

Supply frequency, f = 50.00 Hz

P, No. of poles = 6

N_s, speed =120f/P = 1000 rpm

$\omega_s = 2\pi N_s$ 6283.2 rad/s

Actual speed, $N_r = (1\text{-}s)N_s = 980$ rpm

P_{g1}, Gap power = $3I_2^2.R_2/s = 3I_1^2.R_f = 5739.4$ W

Internal Mech. Power = $(1\text{-}s)P_{g1} = 5624.6$ W

Friction + Windage + core loss = 403.0 W

Output power = Int.Mech.Power –friction –Windage – Core loss = 5221.6 W

Output torque = Output power × $60/2\pi N_r$ = 50.91 N-m

Rotor Cu loss = slip × P_{g1} = 114.79 W

Stator loss = $3.I_1^2.R_1$ = 311.01 W

$$\text{Efficiency} = \frac{\text{output}}{\text{output} + \text{losses}} = 0.86$$ **Answer**

Prob.3.2.2 Induction Motor-Maximum Torque and Starting Torque

The rotor resistance and reactance of a 4-pole, 50 Hz, 3-phase slip-ring induction motor is 0.4 and 4 Ω/phase respectively. Calculate at what speed the torque is a maximum and the ratio of the maximum torque to the starting torque.

What value must the external resistance per phase have so that the starting torque is half of the maximum torque?

$R_2 = 0.4\ \Omega$

$X_2 = 4\ \Omega$

Slip at max. torque = $R_2/X_2 = 0.1$

Supply frequency, f = 50 Hz

P, No. of poles = 4

Ns, speed =120f/P = 1500 rpm

Nr, actual speed = N_s (1-s) = 1350 rpm **Answer**

Since, s = 1 at starting,

$$\frac{T_{start}}{T_{max}} = \frac{R_2/(R_2^2 + X_2^2)}{sR_2/(R_2^2 + s^2X_2^2)} = 0.198$$ **Answer**

$$\frac{sR_2}{R_2^2 + s^2X_2^2} = 0.125$$

Therefore, T_{max} = Constant × 0.125

Starting torque can be made equal to half the max. torque, by introducing an external resistance r in the rotor circuit.

$$T_{start} = 0.5 \times T_{max} = \text{Constant} \times 0.0625 = \frac{\text{Constant}\,(R_2 + r)}{(R_2 + r)^2 + X_2^2}$$

or,

$$0.0625r^2 + r\,(0.125R_2 - 1) + 0.0625R_2^2 + 0.0625X_2^2 - R_2 = 0$$

This is a quadratic equation of the form :

$ar^2 + br + c = 0$ where,

a = 0.0625

b = – 0.95

c = 0.61

Therefore,

$$\frac{-b + \sqrt{b^2 - 4ac}}{2a} = 14.528\ \Omega \quad \text{High value}$$ **Answer**

$$-\frac{-b - \sqrt{b^2 - 4ac}}{2a} = 0.6718\ \Omega \text{ Low value}$$

Higher value is preferable.

Prob.3.2.3 Induction motor starting

The rotor of a 3-phase induction motor is star-connected and has an induced e.m.f. of 50 V between slip-rings at standstill on open circuit, when the stator is connected to a normal voltage supply. The impedance of the rotor at standstill is (0.5 + j3.5) Ω/phase.

Calculate the current in each phase and the power factor at the instant of starting when (a) the rotor is connected to a star -connected external resistance of 4 Ω/phase, (b) when the slips are short-circuited.

Solution.

(*a*) R_2 = rotor resistance = 0.5 Ω

X_2 = rotor reactance = 3.5 Ω

r = external resistance = 4 Ω

$\sqrt{(R_2+r)^2+X_2^2} = 5.700\ \Omega$

I, current = V/Z = 5.063 Amps **Answer**

p.f. = $(R_2+r)/Z$ = 0.789 lag

(*b*) $\sqrt{R_2^2+X_2^2} = 3.535\ \Omega,\ V = 50/\sqrt{3}$

I = V/Z = 8.165 Amps **Answer**

p.f. = R_2/Z = 0.141 (poor starting torque)

Prob.3.2.4 Induction Motor- Testing and Equivalent Circuit

A 12 -pole, 420 V, 10 kW, 50 Hz, three-phase induction motor yielded the following test results:

Open-circuit test: *420V, 6.7 A, 500 W (including 230 W mechanical loss)*

Locked -rotor test: *99V, 14 A, 980 W*

Calculate the parameters of the equivalent circuit per phase with an assumed star-connected stator winding, assigning the copper losses and leakage reactance volt-amperes, equally between stator and rotor windings .

Determine also the torque, mechanical output power, input line current, power factor, and efficiency if the machine operates with a slip of 0.03, using the approximate equivalent circuit. If the stator phase windings are actually connected in delta, estimate the value of resistance per phase, likely to result from a direct measurement.

Solution.

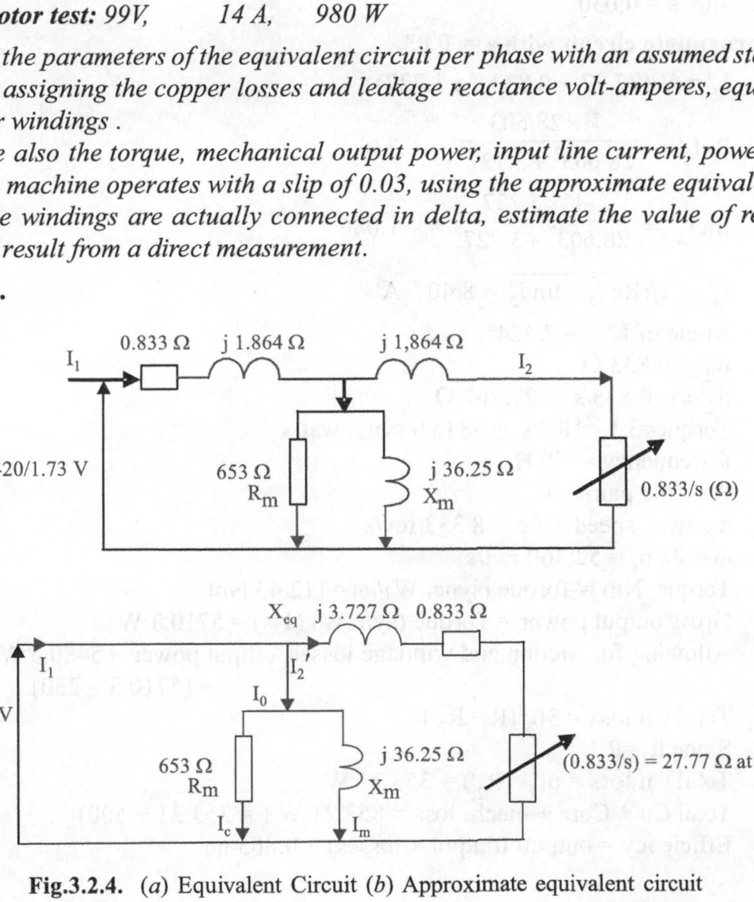

Fig.3.2.4. (*a*) Equivalent Circuit (*b*) Approximate equivalent circuit

From open circuit test data:

V, voltage / phase = 242.49 V
OC test loss = 500 W
Mech. Loss = 230 W

Therefore,

Core loss = P = 270 W
Core loss/ph = 90 W
$R_m = V^2/P$ = 653.33 Ω
$I_c = V/R_m$ = 0.371 A
I_o, No-load current = 6.700 A
$I_m = \sqrt{I_0^2 - I_C^2}$ = 6.690 A
$X_m = V/I_m$ = 36.248 Ω

From short-circuit test data:

V', applied voltage per phase = 99/1.73 = 57.225 V
I', short-circuit current = 14.0 A
$Z_{eq} = V'/I'$ = 4.088 Ω
P', copper loss = 980 W

$R_{eq} = \dfrac{P'}{3I'^2}$ = 1.667 Ω (= sum of stator and rotor resistances, $R_1 + R_2$)

$X_{eq} = \sqrt{Z_{eq}^2 - R_{eq}^e}$ = 3.732 Ω

slip, s = 0.030

From the approximate circuit with s = 0.03,

$I_2' = V/(27.77 + 0.833 + j\,3.727)$

$\text{Re}I_2' = \dfrac{V \times 28.603}{28.603^2 + 3.727^2} = 8.336$

$\text{Im}\,I_2' = \dfrac{-V \times 3.727}{28.603^2 + 3.727^2} = -1.086$

$I_2' = \sqrt{(\text{Re}\,I_2' + \text{Im}\,I_2'} = 8.407$ A

Angle of $I_2' = -7.424°$.
$R_2' = 0.833$ Ω
$R_2'/s = 0.833/s = 27.767$ Ω
Torque=3 $I_2'^2 * R_2'/s$ = 5886.9 sync. watts
f, frequency = 50 Hz
P = Pole pairs = 6
n_s, sync. speed = f/p = 8.333 rev/s
$\omega = 2\pi\, n_s$ = 52.360 rad/s
Torque, Nm = Torque (sync. W)/ ω =112.43 Nm **Answer**
Gross output power = Torque (syn. W) (1-s) = 5710.3 W
Allowing for friction and windage losses, output power = 5480.3 W
= (5710.3 – 230) **Answer**

Total Cu loss = $3I_2^2(R_1+R_2')$
Since $R_1 = R_2'$,
Total Cu loss = $6I_2'^2(R_2')$ = 353.21 W
Total Cu + Core + mech. loss = 853.21 W (= 353.21 + 500)
Efficiency = output/ (output + losses) = 0.865 pu **Answer**

Input line current, $I_1 = I_2'+I_o$
$R_eI_1 = I_c + Re\,I_2' = 8.707$
$ImI_1 = -Im + ImI_2' = -7.776$
$I_1 = \sqrt{(Re\,I_1)^2 + (Im\,I_1)^2} = 11.674$ A **Answer**
Input p.f. = 0.746 lag **Answer**

$R_1 = 0.833\ \Omega$ is the stator resistance per phase of equivalent star. For the same power developed in delta with the same line voltage and current, impedance per phase must be increased by a factor of 3. Due to skin effect, resistance under a.c. conditions exceeds that for d.c. so the value of stator phase resistance will be somewhat less than 2.5 Ω.

Prob.3.2.5 Induction Motor-Starting current, Starting Torque & Max. Torque

For the 11.19 kW, 4-pole motor described below, find the slip at which maximum torque occurs, maximum torque ,slip at which maximum power occurs, maximum power ,starting torque, and starting current.

220 V(L-L), 3-phase, 11.19k W, 1725 r.p.m. $R_1 = 0.15\ \Omega$, $R_2 = 0.2\ \Omega$, $X_1 = X_2 = 0.3\ \Omega$, *Friction & windage loss = 300 W, frequency = 60 Hz*

Solution.

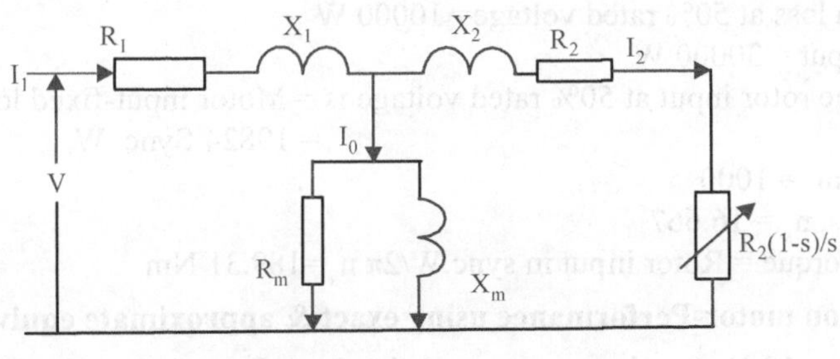

Fig.3.2.5. *Equivalent Circuit*

V / phase = 127.02 V
$R_1 = 0.15\ \Omega$
$R_2 = 0.2\ \Omega$
$X_1 = X_2 = 0.3\ \Omega$

Slip at max. torque = $\dfrac{R_2}{\sqrt{R_1^2 + (X_1+X_2)^2}} = 0.323$ **Answer**

$T_{max} = \dfrac{3V^2}{2[R_1 + \sqrt{R_1^2 + (X_1+X_2)^2}\,]} = 31491$ sync. W

P, Pole pairs = 2
f, frequency = 60 Hz
sync. speed, $n_s = f/P = 30$
T_{max}, Nm = T_{max} (sync. W)/2π n_s = 167.07 Nm **Answer**

$Z_{eq} = \sqrt{(R_1+R_2)^2 + (X_1+X_2)^2} = 0.694\ \Omega$
Slip at max. power = $R_2/(R_2 + Z_{eq}) = 0.223$ **Answer**

Max. Power = $\dfrac{3V^2}{2(R_2 + Z_{eq})} = 27051$ W **Answer**

Starting torque in synchronous watts = $\dfrac{3V^2R_2}{Z_{eq}^{\ 2}} = 20062$ W

Starting torque = Torque in sync.W/2π n_s = 106.43 Nm **Answer**
Starting current = V/Z_{eq} = 182.86 A **Answer**

Prob.3.2.6 Induction Motor: Starting torque from Tests

A 3-phase 14.92 kW 400V, 20 A, 1000 rpm squirrel -cage motor drives a pump. Calculate the starting torque of the motor at 50 % rated voltage from the following test information:

No-load: 440 V,10A,1000W,1100 rpm

Blocked rotor: 200V,80A,20kW

Resistance between stator terminals = 1 Ω

At 50% rated voltage, the current is 100 A and the motor input is 30 kW. The stator copper loss is 10 kW.

Solution: No load loss = 1000 W

Stator resistance /phase, r = 0.5 Ω

I, Stator current at no load = 10 A

Stator cu loss at no load = $3I^2r$ = 150 W

Fixed losses = No-load loss – Stator cu loss at no load = 850 W

Starting voltage = 200 V

Test voltage = 440 V

Fixed loss modified by the square of the ratio of starting voltage to the test voltage = (Starting voltage / Test voltage)^2 = 175.62 W

Stator Cu loss at 50% rated voltage= 10000 W

Motor input = 30000 W

Hence, the rotor input at 50% rated voltage is = Motor input-fixed loss – Stator cu loss = 19824 Sync. W

speed, rpm = 1000

speed, rps, n_s = 16.667

Starting torque = Rotor input in sync.W/2π n_s =189.31 Nm **Answer**

Prob.3.2.7. Induction motor-Performance using exact & approximate equivalent circuits

A 400 V 3-phase,50 Hz 8 pole star-connected induction motor has the following equivalent circuit parameters:

$R_1 = 0.1\ \Omega, R_2 = 0.3\ \Omega, X_1 = 0.6\ \Omega, X_2 = 2\ \Omega$

Magnetizing branch admittance = Y_m = 0.005 –j0.05 mho referred to primary side. Full-load slip = 5%.

(a) Determine the full-load gross torque, stator input current and power factor using both approximate & exact equivalent circuits. The effective stator/rotor turns ratio per phase is 1/1.5.

(b) Using the approximate circuit of (a), calculate the output of the machine when driven at a speed of 780 rpm.

Solution. $R_2'/s = 2.667\Omega$ $jX_2' = 0.889\ \Omega$

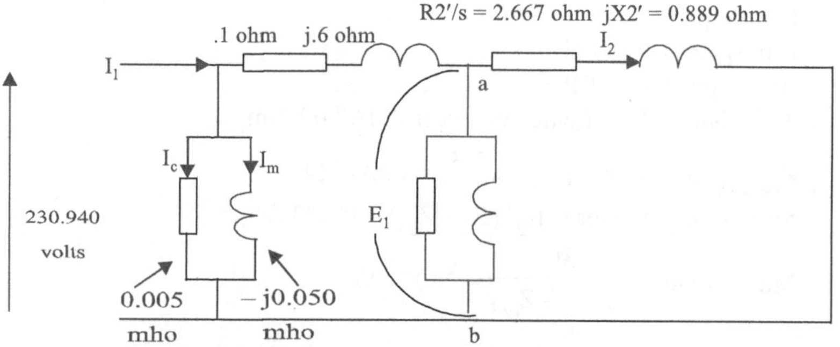

Fig.3.2.7. *Exact equivalent circuit*

Supply volt/phase =V1= 230.94 Volts

T, Stator/Rotor turns = 0.667

$R_1 = 0.100\ \Omega$
$R_2 = 0.300\ \Omega$
$X_1 = 0.600\ \Omega$
$X_2 = 2.000\ \Omega$
$\text{Re}\,Y_m = 0.005\ \mho$
$\text{Im}\,Y_m = -0.050\ \mho$
$\text{Slip} = 0.050$
$R_2' = R_2T^2 = 0.133\ \Omega$
$X_2' = X_2T^2 = 0.889\ \Omega$

Approximate circuit:

$\text{Re}\,Z = R_1 + R_2'/s = 2.767\ \Omega$
$\text{Im}\,Z = X_1 + X_2' = 1.489\ \Omega$
$Z_{mag} = 3.142\Omega$
$I_2' = V_1/Z$

$$\text{Re}I_2' = \frac{V_1 \text{Re}\,Z}{(\text{Re}\,Z)^2 + (\text{Im}\,Z)^2} = 64.727\ \text{A}$$

$$\text{Im}I_2' = -\frac{-V_1 \text{Im}\,Z}{(\text{Re}\,Z)^2 + (\text{Im}\,Z)^2} = -34.83\ \text{A}$$

$I_2'\text{mag} = 73.504\ \text{A}$
$I_c = V_1.\text{Re}Y_m = 1.154\ \text{A}$
$\text{Im} = V_1.\text{Im}\,Y_m = -11.55\ \text{A}$
$\text{Re}I_1 = I_c + \text{Re}I_2' = 65.882\ \text{A}$
$\text{Im}I_1 = \text{Im} + \text{Im}I_2' = -46.38\ \text{A}$
$I_{1mag} = 80.57\ \text{A}$ **Answer**
$f = \text{frequency} = 50\ \text{Hz}$
p, pole pairs $= 4$
n_s, speed $= f/p = 12.5$ rev/s
Torque $= (3 \times I_2'^2 \times R_2'/s)/(2\pi n_s) = 550.33$ Nm **Answer**
Input pf $= \text{Re}I_1/I_1\text{mag} = 0.817$lag **Answer**

Exact equivalent circuit

Admittance between a and b is:

$Y = Y_m + 1/Z$

$$\text{Re}Y = \text{R}_e Y_m + \frac{R_2'/s}{\left(\dfrac{R_1'/s}{s} + X_2'^2\right)} = 0.342\ \mho$$

$$\text{Im}Y = \text{Im}Y_m \frac{X_2'}{\left(\dfrac{R_2'}{s}\right)^2 + X_2'^2} = -0.163\ \mho$$

$Z_{ab} = 1/Y$
$\text{Re}Z_{ab} = \text{Re}\,Y/(\text{Re}\,Y^2 + \text{Im}\,Y^2) = 2.383\ \Omega$
$\text{Im}\,Z_{ab} = -\text{Im}Y/(\text{Re}Y^2 + \text{Im}Y^2) = 1.130\ \Omega$
$Z_{ab}\,\text{mag} = 2.637\ \Omega$
Z = Input impedance
$\text{Re}Z = R_1 + \text{Re}Z_{ab} = 2.483\ \Omega$
$\text{Im}Z = X_1 + \text{Im}\,Z_{ab} = 1.730\Omega$
$Z_{mag} = 3.026\ \Omega$

$$\text{Input current} = V_1/Z_{mag} = 76.297 \text{ A}$$ **Answer**

$$\text{Input p.f.} = \cos\left(\tan^{-1}\frac{\text{Im}\,Z}{\text{Re}Z}\right) = 0.820 \text{ lag}$$ **Answer**

To find the e.m.f. E_1, since the same current I_1 flows through Z and Z_{ab} in series, the voltage across each impedance is proportional to the impedance moduli *i.e*,

$$E_1 = V_1 Z_{ab}/Z = 201.26 \text{ V}$$

Hence,
$$I_2' = \sqrt{E_1^2/\left(\frac{R_2'^2}{s^2}+X_2'^2\right)} = 71.6 \text{ A}$$

Torque is proportional to current squared.

Therefore,

$$\text{Torque} = 550.335\left(\frac{71.599}{73.504}\right)^2 = 522.19 \text{ Nm}$$ **Answer**

$$I_c = E_1 \text{ Re}Y_m = 1.006 \text{ A}$$
$$I_m = E_1 \times \text{Im}Y_m = -10.06 \text{ A}$$
$$\text{Speed, } N_r = 780 \text{ rpm}$$
$$n_r = 13 \text{ rev/s}$$
$$\text{slip} = \frac{n_s - n_r}{n_s} = -0.04$$
$$r_2 = \text{Apparent rotor circuit resistance} = R_2'/\text{slip} = -3.333\ \Omega$$

Apparent impedance, Z',

$$\text{Re}\,Z' = R_1 + r_2 = -3.23\ \Omega$$
$$\text{Im}\,Z' = X_1 + X_2' = 1.48\ \Omega$$
$$\text{Re}I_2' = \frac{V_1 \text{Re}\,Z}{(\text{Re}\,Z)^2+(\text{Im}\,Z)^2} = -59.09 \text{ A}$$
$$\text{Im}I_2' = \frac{-V_1.\text{Im}\,Z}{(\text{Re}\,Z)^2+(\text{Im}\,Z)^2} = -27.08 \text{ A}$$

From the approx. circuit:

$$I_o = 1.15 - j11.5.$$

Therefore,

$$\text{Re}I_1 = \text{Re}I_2' + 1.15 = -57.94 \text{ A}$$
$$\text{Im}I_1 = \text{Im}I_2' - 11.5 = -38.58 \text{ A}$$
$$I_{1mag} = 69.609 \text{ A}$$
$$\text{Output kVA} = 1.73 \times 400 \times I_{1mag} = 48.168 \text{ kVA}$$ **Answer**
$$\text{p.f.} = \text{Re}I_1/I_{1mag} = -0.832 \text{ lead}$$

Problem 3.2.8 Induction Motor- Efficiency and torque

(*i*) *An induction motor draws 25 A from 460 V three-phase line at a power factor of 0.85 lagging. The stator cu loss is 1000W, and the rotor cu loss is 500 W.*

Rotational, windage, and friction losses = 250 W, Core loss = 800 W, stray load loss = 200 W. Calculate

(a) the air gap power, P_g
(b) the developed mechanical power
(c) output horse power
(d) efficiency

(*ii*) *If the frequency of the source in part (i) is 50 Hz, and the machine has four poles find*
(a) slip, (b) operating speed (c) developed torque (d) output torque

Solution. (i)

(a) V, Supply voltage (L-L) = 460 V

I_1, Input current = 25 A

Stator Cu loss = 1000 W

Input pf = 0.85 lag

Input power = $1.73 \times VI_1 \times$ p.f. = 16931 W

P_g, air gap power = Input power - Stator loss = 15931 W

(b) Rotor cu loss = 500 W

Developed mech. power = P_g - Rotor cu loss = 15431 W

(c) Core loss = 800 W

Friction + windage = 250 W

Stray load loss = 200 W

Output power = Developed power- Core friction - Windage-Stray load losses =14181W

Efficiency = Output/Input = 0.837

(ii)

(a) Slip = Rotor Cu loss /P_g = 0.0314 **Answer**

f, frequency = 50

p, pole pairs = 2

n_s, speed = f/p = 25 rev/s

$n_r = n_s(1\text{-slip})$ = 24.215 rev/s

$N_r = n_r \times 60$ = 1452.9 rpm **Answer**

Developed torque = Developed mech power/$2\pi\, n_r$ = 101.47 Nm **Answer**

Output torque = Output power/$2\pi\, n_r$ = 93.203 Nm. **Answer**

Prob.3.2.9. Induction motor Starting- with autotransformer, star-delta switch, stator resistance starter

The full load speed of a 4-pole, 50 Hz 3-phase squirrel -cage motor is 1440 rpm. The ratio of starting current to full-load current is 5. Calculate the ratio of the starting torque to full-load torque, in the following methods of starting:

(a) auto-transformer with 60 % tapping,

(b) star-delta switch, and

(c) by a stator resistor starter to give a starting current of twice the full - load current.

Solution. f, frequency 50 Hz

p, pole pairs 2

n_s syn. speed 25

$N_{r,}$ speed 1440 rpm

$n_{r,}$ speed 24 rev/s

s, slip = $\dfrac{n_s - n_r}{n_s}$ = 0.04

Torque (T) $\alpha\ I_2^2/s$.

I_s is the Starting current

I_{fl} is the Full load current

$T_{start}/T_{full\ load} = I_s^2.s/I_{fl}^2$

(a) Auto Transformer

$$I_s = 0.6 \times 5I_{fl} = 3I_{fl}$$

$$I_s/I_{fl} = 3$$

$$T_{start}/T_{fl} = (I_s/I_{fl})^2 \times s = 0.36$$ **Answer**

(b) Star-Delta Switch

Voltage impressed per phase during starting = V/1.73205

Therefore, starting current/Full load current = 5/1.73, or

$I_s/I_{fl} = 2.8902$

$T_{start}/T_{fl} = (I_s/I_{fl})^2 *$ slip= 0.334 **Answer**

(c) Stator –resistor starter

$I_s/I_{fl} = 2$

$T_{start}/T_{fl} = (I_s/I_{fl})^2$ slip= 0.16 **Answer**

Prob.3.2.10 Induction Motor- Single-phase -Capacitor start

A 0.746 kW, 120 V, 60 Hz, capacitor start motor has the following constants for the main and auxiliary windings (at starting) :

Main winding, $Z_m = 4.5 + j\,3.7\ \Omega$

Auxiliary winding, $Z_a = 9.5 + j\,3.5\ \Omega$

Find the value of starting capacitance that will place the main and auxiliary winding currents in quadrature at starting.

Solution.

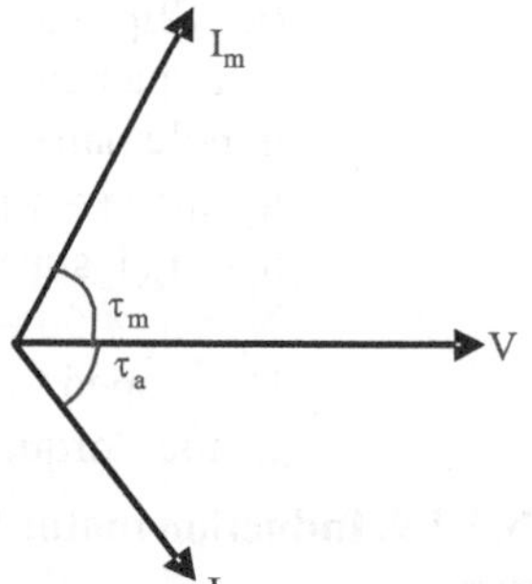

Fig.3.2.10. Phasor diagram

$R_m = 4.5\ \Omega$

$X_m = 3.7\ \Omega$

$R_a = 9.5\ \Omega$

$X_a = 3.5\ \Omega$

$\tau_m = \tan^{-1}(X_m/R_m) = 0.6881$ radian = 39.427°

$\tau_a = \tau_m - (\pi/2) = -0.883$ radian = – 50.572°

X_c = Reactance of capacitor

$$\tau_a = \tan^{-1}\left(\frac{X_a - X_c}{R_a}\right)$$

Therefore,

$X_c = X_a - R_a.\tan\tau_a = 15.054\ \Omega$

f = 60 Hz

$$\text{Capacitance, C} = \frac{I}{2\pi f.X_c} = 0.000176\ \text{F}$$

$= 176\ \mu F$ **Answer**

Prob.3.2.11 Induction Motor-Calculation of Rotor Resistance for Reduction in Speed

A 4-pole ,50 Hz, 3-phase slip ring induction motor when fully loaded runs with a slip of 3%. Calculate the value of the resistance necessary to reduce the speed by 10 %. Each rotor phase has a resistance of 0.2 Ω.

Solution.

f Supply frequency = 50 Hz.

P, No. of poles = 4

N_s, speed = 120f/P = 1500 rpm

N_r = Actual speed

s = slip = 0.03 pu

$N_r = N_s(1 - s) = 1455$ rpm

N_r' New speed $= N_r - 0.1\,N_r = 1309.5$ rpm

s′ New slip $= (N_s - N_r')/N_s = 0.127$

=12.7% **Answer**

R and R' are the total resistances per phase in the rotor at slips s and s' respectively.

Therefore,

$$R' = Rs'/s$$
$$R = 0.2\ \Omega$$
$$R' = 0.846\ \Omega$$

External resistance / phase = $R' - R = 0.646\ \Omega$ **Answer**

Prob.3.2.12 Single-phase Induction motor-Performance

A single-phase induction motor is rated 186.5 W, 100 V, 50 Hz, 4 poles. The equivalent circuit parameters are:

$R_1 = 2\ \Omega,\ R_2' = 4\ \Omega,\ X_m = 60\ \Omega,\ X_1 = X_2' = 2\ \Omega$

Rotational loss (Core + Friction + Windage) = 40 W

Rotor is operating at rated voltage and frequency.

Slip = 0.05

Find the input current and the resultant electro-magnetic torque.

Solution :

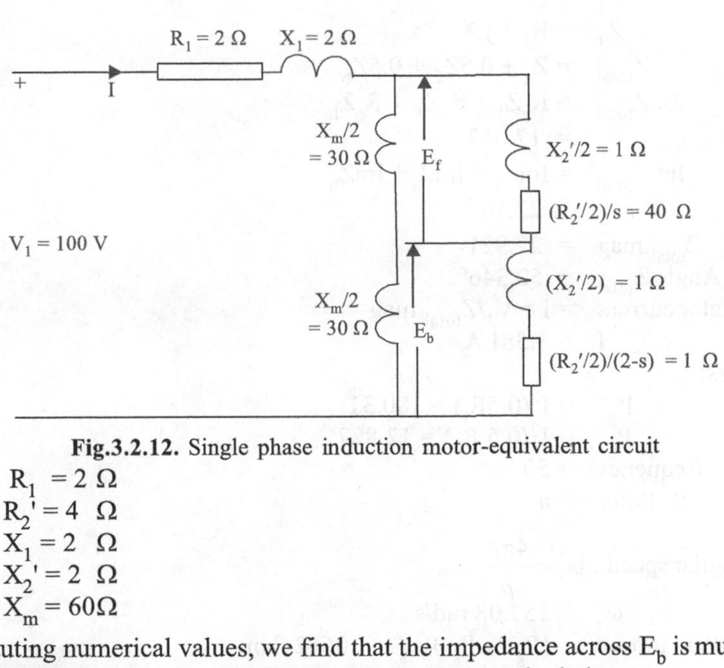

Fig.3.2.12. Single phase induction motor-equivalent circuit

$R_1 = 2\ \Omega$

$R_2' = 4\ \Omega$

$X_1 = 2\ \Omega$

$X_2' = 2\ \Omega$

$X_m = 60\Omega$

On substituting numerical values, we find that the impedance across E_b is much smaller than that across E_f. Therefore, great accuracy is not required in its calculation.

Thus,

$$0.5\ Z_b = \frac{1}{2}\frac{jX_m\left[R_1'/(2-s) + jX_2'\right]}{R_2'/(2-s) + j(X_m + X_2')}$$

or,

$$0.5\ Z_b \text{ is approx.} = 0.5\left[\frac{R_2}{2} + jX_2'\right]$$

$$R_e(0.5Z_b) = 0.5R_b = 1$$
$$Im(0.5Z_b) = 0.5X_b = 1$$

$$0.5Z_f = \frac{\frac{jX_M}{2}\cdot\left(\frac{R_2'/2}{s}+\frac{jX_2'}{2}\right)}{\frac{R_2'}{s}+\frac{jX_M}{2}+j\frac{X_2'}{2}}$$

$$0.5Z_f = \frac{j30\,(40+j1)}{40+j31}$$

$\text{Re}\,(0.5Z_f) = 0.5R_f = \text{Re}(j30 \times (40 + j1) \times (40 - j31)/(1600 + 31 \times 31) = 36000/(1600 + 31 \times 31)$

$0.5R_f = 14.057$

$I_m(0.5Z_f) = 48930(1600+31^2)$

$= 19.106$

$R_1 = 2$

$X_1 = 2$

$Z_1 = R_1 + j\,X_1$

$Z_{total} = Z_1 + 0.5Z_f + 0.5Z_b$

$\text{Re}\,Z_{total} = R_eZ_1 + R_eZ_f + R_eZ_b$

$= 17.057$

$\text{Im}\,Z_{total} = \text{Im}Z_1 + \text{Im}Z_f + \text{Im}Z_b$

$= 22.106$

$Z_{total}.\text{mag} = 27.921$

Angle $Z_{total} = 52.346°$.

Stator current $= I = V_1/Z_{total}.\text{mag}$

$I = 3.581$ A **Answer**

Gap powers:

$P_g^+ = I^2(0.5R_f) = 180.31$

$P_g^- = I^2(0.5\,R_b) = 12.827$

f, frequency = 50

P, Poles = 4

Synchronous angular speed, $\omega_s = \frac{4\pi f}{p}$

$\omega_s = 157.08$ rad/s

Resultant torque $= (P_g^+ - P_g^-)/\omega_s = 1.0662$ Nm **Answer**

Prob.3.2.13 Induction motor performance

A 4-pole 400 V, 3-phase, 50 Hz induction motor runs at 1440 rpm at 0.88 p.f. lagging and delivers 10.817 kW. The stator loss is 1060 W, and friction & windage losses are 375 W. Calculate

(i) slip
(ii) rotor Cu loss
(iii) rotor frequency
(iv) line current
(v) efficiency

Solution. f, Supply frequency = 50 Hz

P, No. of poles = 4

N_s, speed = 120f/P = 1500 rpm

N_r = Actual speed = 1440 RPM

s = slip 0.04 pu **Answer**

Motor output = 10817 W

Stator Cu loss = 1060 W

Friction & windage = 375 W

Rotor Cu loss = Rotor input-motor output-friction & Windage

Rotor input = Rotor Cu loss/slip

Therefore,

Rotor cu loss = (– motor output-friction & windage) (s/(s – 1)) = 466.33W **Answer**

Rotor input = rotor Cu loss/slip = 11658 W

Motor input = rotor input + stator loss =12718 W

V = 400 Volts

Cos ϕ = 0.88 lag

Line Current = Motor input/(1.73 V CosΦ) = 20.885 A **Answer**

$$\text{Efficiency} = \frac{\text{Output}}{\text{Input}} = 0.850 \text{ pu} = 85\%$$ **Answer**

Rotor e.m.f. frequency = s × f = 2 Hz **Answer**

Prob.3.2.14 Induction Motor -Maximum torque & starting current with normal & reduced voltages

A 3-phase, 6 pole, 50 Hz induction motor has a peak torque of 6 Nm and a starting torque of 3 Nm when operating at full voltage. Maximum torque occurs at a slip of 25%.When started at 1/3 rd of normal voltage the current is 2 A.

(a) *Find the mechanical power at peak torque when operating at normal voltage*

(b) *Find the maximum torque at 1/3 rd of normal voltage*

(c) *Find the starting current with 1/3 rd of normal voltage*

(d) *Find the extra rotor circuit resistance as a percentage to give maximum torque at starting. What would then be the current in terms of that at peak torque without external resistance ?*

Solution.

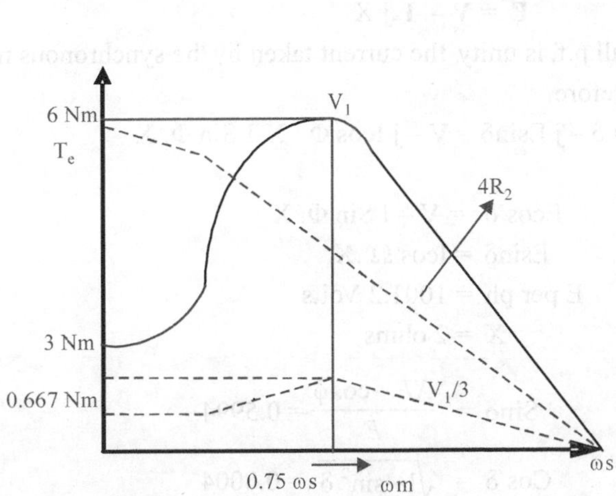

Fig.3.2.14. *Torque-Speed Curve*

f, frequency = 50 Hz

p, Pole pairs = 3

S, slip = 0.25 pu

T_e Max. Torque = 6 Nm

(a) Power at max torque = $\omega_m.T_e = 2\pi n_r T_e = 2\pi n_s (1\text{-}s) T_e = 2\pi f (1\text{-}s) T_e/p = 471.24$ W

Answer

(b) T α V^2

Therefore reduced max. torque at (1/n) of V = T_e/n^2

Here,

n = 3

Reduced max. torque = 0.666 Nm

(c) I α V

$I_{starting}$ at normal voltage = $I_{starting}$ at reduced voltage × normal voltage/ fraction of reduced voltage = 2 (n) =6A

Answer

(d) A given torque requires a particular value of R_2'/s. Since, s changes from 0.25 to 1, then the total circuit resistance must change in the same ratio *i.e.* by 4 times.

Hence, extra rotor resistance = 300% R_2'

Since, R_2'/s is constant and since this is the only equivalent circuit impedance which could vary with speed , the total impedance presented to the terminals is unchanged.

So, the current is the same as at *s* = 0.25 pu

Prob.3.2.15 Induction Motor in parallel with Synchronous Motor

A 3-phase induction motor and a 3-phase star-connected synchronous motor are both operating off a common 3-phase 1039 V (L-L) supply, the combined power factor of the two motors being unity and each motor taking an input of 540 kW. The synchronous motor has negligible resistance, a synchronous reactance of 2 Ω / phase and an induced e.m.f. of 1732 kV (L-L). Find the power factor of each motor and draw the phasor diagram of the synchronous motor.

Solution. Synchronous motor

P, Power / ph = 180000 W

V per ph = 599.87 Volts

Icos Φ = P/V= 300.07

$$\mathbf{E} = \mathbf{V} - \mathbf{I}.jX$$

Since the overall p.f. is unity, the current taken by the synchronous motor must be leading.

Therefore,

$$E\cos\delta - j\,E\sin\delta = V - j\,I\cos\Phi\,.X - I\sin\Phi.\,X$$

or,

$$E\cos\delta = V - I\sin\Phi.\,X$$

$$E\sin\delta = I\cos\Omega\,.X$$

E per ph = 1001.2 Volts

X = 2 ohms

$$\sin\delta = \frac{XI - \cos\phi}{E} = 0.5994$$

$$\cos\delta = \sqrt{1-\sin^2\delta} = 0.8004$$

$$I\sin\Phi = (V - E\cos\delta)/X = -100.7$$

$$I\cos\Phi = E\sin\delta/X = 300.07$$

$$I = 316.53\text{ A}$$

$$\cos\Phi = 0.948 \text{ lead}$$

Answer

Power diagram:

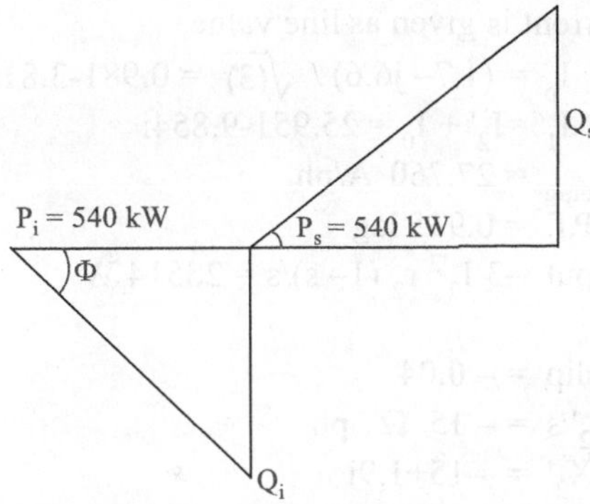

Fig.3.2.15. *Power diagram of the two motors*

P_i, Q_i refer to the induction motor and P_s ,Q_s refer to the synchronous motor.

Since the overall p.f. is unity, the induction motor p_f must be the same , but lagging.

Prob.3.2.16. Induction motor driven as a generator

The following data refer to a 415 V, 50 Hz 3-phase delta-connected 6-pole induction machine. stator leakage impedance = 0.7 + j1.9 Ω / phase, rotor standstil leakage impedance (referred to stator) = 0.6 + j1.9 Ω /phase, no-load current = 1.7 + j 6.6 A/line.

Determine,

(i) the current, power factor and mechanical power output when running as a motor at a slip of 0.04.

(ii) the power output and power factor when driven as a generator at a speed of 1.04 times synchronous speed.

(iii) the initial braking torque when, with the motor running at a slip of 0.04, two of the stator line connections are suddenly interchanged.

Solution.

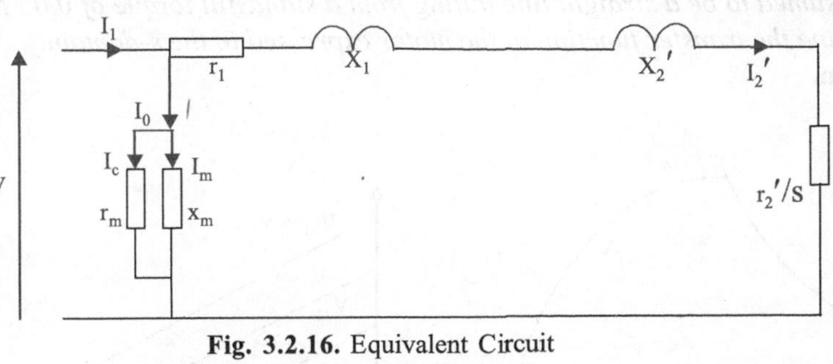

Fig. 3.2.16. Equivalent Circuit

$$r_1 = 0.7\ \Omega$$
$$X_1 = 1.9\ \Omega$$
$$r' = 0.6\ \Omega$$
$$X_2' = 1.9\ \Omega$$
$$I_o = 1.7 + j6.6\ A$$

Part (i)

$$s = 0.04$$
$$r_2'/s = 15$$
$$V = 415$$
$$r_1 + jX_1 = 0.7 + 1.9i,$$

$r_2'/s + jX_2' = 15 + 1.9i$, $I_2' = V/(r_1 + jX_1 + r'_2/s + jX_2) = 24.970 - j6.043$

$I_2' = (\text{mag}) = 25.69$

Since no-load current is given as line value,

$I_o = (1.7 - j6.6) / \sqrt{(3)} = 0.981\text{-}3.810i$

$I_1 = I_2' + I_0 = 25.951\text{-}9.854i$

$I_{1mag} = 27.760$ A/ph

P.f. = 0.935 lag

Total output $= 3\, I_2'^2\, r_2\, (1 - s)/s = 28514$ W

Part (ii)

slip = – 0.04

$r_2'/s = -15\ \Omega$ / ph

$r_2'/s + jX_2' = -15+1.9i$

$I_2' = V/(r_1 + jX_1 + r_2'/s + jX_2') = -27.106\text{-}7.203i$

$I_1 = I_o + I_2'$ (A/phase) $= -26.125 - 11.013i$

$I_{1mag} = 28.352$ A/ph

p.f. = 0.921 lead

Power delivered $= 3VI_1 \cos\pi = 32526$ W

Part (iii)

Direction of field is reversed, so slip = 2–s = 1.96

$r_2'/s = 0.306$

$r_2'/s + jX_2' = 0.306 + 1.9i$

$I_2' = 27.021 - 102.056i$

$I_{2\,mag}'$ 105.573 A/ph

Torque = rotor input/w $= 3I_2'^2 r'_2/s = 98.421$ N–m

Prob. 3.2.17 Two phase servo motor

A two-phase servo motor has similar windings rated at 110 V. The motor inertia is 1.8×10^{-6} kg.m^2 and the friction torque is 70×10^{-6} Nm s /rad. The torque /speed curve at full control voltage is assumed to be a straight line falling from a standstill torque of 0.05 Nm to zero at 400 rpm. Determine the transfer function of the motor expressed in the s-domain.

Solution.

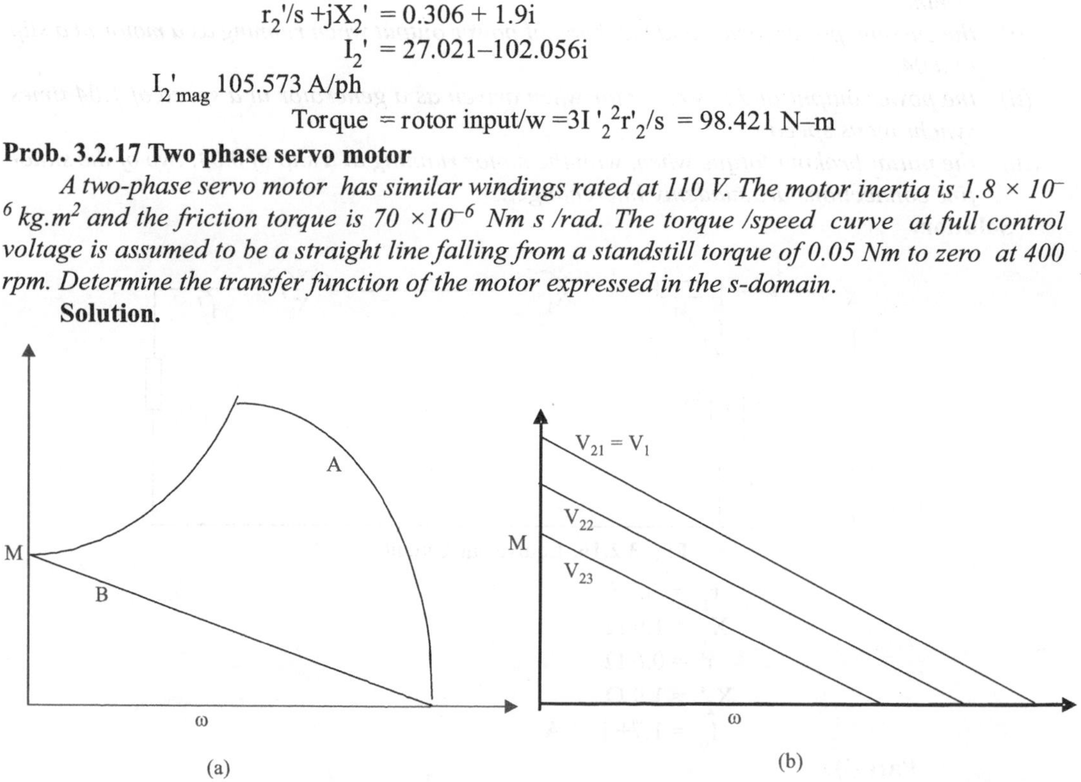

Fig.3.2.17. *Speed-torque curves for 2-phase servomotor*

Torque-speed curve as in Fig.3.2.17(b):

$$M_e = kV_2 - m\frac{d\theta}{dt}$$

Mechanical torque $M_m = \frac{Jd^2\theta}{dt^2} + f\frac{d\theta}{dt}$

Equating and transferring to s domain,

$KV_2(s) - ms\theta(s) = Js^2\,\theta(s) + fs\,\theta(s)$

$$\text{Transfer function} = \frac{\theta(s)}{V_2(s)} = \frac{K}{Ms + Js^2 + fs} = \frac{K}{s(f+m) + Js^2}$$

$$= \frac{K}{f+m}\left(\frac{1}{s\left(1+\left(\frac{J}{f+m}\right)\right)}\right)$$

K = 0.05/110 = 0.000455 Nm/V

ω_0 at zero torque = 4000 × 2 × π/60 = 418.88 rad/s

m = M_0/ω_0 = 0.05/ω_0 = 0.000119 Nm–s/rad

f = 0.00007 Nm s/rad

f + m = 0.000189

K/(f + m) = 2.400

0.0000018/0.000189 = 0.009

Transfer function = 2.4s (1+ 1.8 × 0.000001s / 0.000189)

$$\frac{\theta(s)}{V(s)} = \frac{2.4}{s(1+0.0095s)}$$

CHAPTER 4

OVERHEAD LINES

Simulator for transmission thermal limits can be found at the website: **http://powerlearn.ee.iastate.edu**. The website **http://www.utexas.edu/~grady/TOWERABC.html/** calculates the line constants for overhead three-phase, single-circuit or double-circuit, transmission lines and produces contour plots of r.m.s. V, E, and H, plus sound. It can be used for 50/60 Hz, and also for harmonic frequencies.

4.1.1 Factors affecting Line design

- Voltage level
- Conductor type & size
- Line regulation & voltage control
- Corona & losses
- Proper load flow & system stability
- System protection
- Insulation co-ordination
- Right of way
- Mechanical design
- Sag & stress calculation
- Conductor composition
- Conductor spacing
- Insulator /conductor hardware selection
- Structural design
- Structure types
- Stress calculations

4.1.2 Conductor size

Gauge sizes decrease as the wire increases in size.

Number of strands = $3n^2 - 3n + 1$

where n = number of layers including the single central strand.

The following conductors are used.

A A C-all aluminium conductor

A A AC-all aluminium alloy conductor

A C S R-aluminium conductor steel re-inforced

A C A R-aluminium conductor alloy re-inforced

4.1.3 Line resistance

$R = \rho l / A$

$R_2/R_1 = (T_0 + T_2)/(T_0 + T_1)$

R_2 = Resistance at temperature, T_2

R_1 = Resistance at temperature, T_1

T_0 = Constant

= 234.5 for annealed copper of 100% conductivity
= 241 for hard drawn copper of 97.3% conductivity
= 228 for hard drawn aluminium of 61% conductivity
Skin effect is function of conductor size, frequency and resistance of conductor material.
Discuss the proximity effect, stranding and spiralling of conductors

4.1.4 Line inductance - one phase & 3-phase

Single-phase overhead line

Voltage drop in a single-phase line due to loop impedance

$$2l\left[(R+\frac{j\omega\mu_0}{2\pi}In\left(\frac{D_m}{Ds}\right)\right]I$$

l = line length, m
R = resistance of each conductor, m
D_m = equivalent or geometric mean distance (GMD) between conductor centres
D_s = Geometric mean radius (GMR), or self-GMD of one conductor
= 0.7788 r for cylindrical conductor
r = conductor radius
I = current
$L = 2 \times 10^{-7} \ln (D_m/D_s)$ H/m

Three-phase overhead line (unsymmetrical spacing)

$D_{ab} \neq D_{bc} \neq D_{ca}$
Equivalent equilateral spacing $= D_{eq} = D_m = (D_{ab} D_{bc} D_{ca})^{1/3}$
In practice, conductors are transposed.
Transposition is carried out at switching stations
Average inductance per phase
$L = 2 \times 10^{-7} \ln (D_{eq}/D_s)$ H/m

4.1.5 Line capacitance, 1-phase & 3-phase

Single-phase overhead line

$C_{ab} = 2\pi\varepsilon_0\varepsilon_r / \ln (D/r)$ (F/m)
The capacitance to neutral for a two- wire line is twice the line-to-line capacitance, Cab.

Three-phase overhead line

Line-to-neutral capacitance
$C_n = 2\pi\, \varepsilon_0 \varepsilon r / \ln (D_{eq}/r)$ (F/m)
Charging current /phase = $j\omega C_n V_{ph}$ (A/m)

4.1.6 Effect of ground on capacitance of 3-phase line

The capacitance of a 3-phase transposed line considering ground effect is given by
$C_n = 2\pi\, \varepsilon_0 \varepsilon_r / [\ln (D_{eq}/r) - \ln (h_{12}\, h_{23}\, h_{31}/h_{11} h_{22} h_{33})]$ (F/m)
where h_{12}= distance between conductor 1 and image of conductor 2, etc. Effect of ground is to increase the capacitance.

4.1.7 Equivalent Circuit for Short Transmission Line (up to 80 km)

Note that bold symbols indicate complex quantities.
$\mathbf{V_s} = \mathbf{V_r} + \mathbf{I_r Z}$
$\mathbf{I_s} = \mathbf{I_r} = \mathbf{I}$
Draw a phasor diagram for a short line with inductive load and with capacitive load, using $\mathbf{V_r}$ as the reference phasor.
Show that

$$Vs = \sqrt{[(V_r + IR\cos\phi_f + (\text{or}-) IX\sin\phi_r)^2 + (IX\cos\phi_r + (\text{or}-) IR\sin\phi_r)^2]}$$

+ sign above is for lagging p.f.
– sign above is for leading p.f
ϕ_r = angle between $\mathbf{V_r}$ & $\mathbf{I_r}$
ϕ_s = angle between $\mathbf{V_s}$ & $\mathbf{I_s}$
$\delta = \phi_s - \phi_r$ = load angle
$\tan\delta = (IX \cos\phi_r + (or -) IR \sin\phi_r) / (V_r + IR \cos\phi_r + (or -) IX \sin\phi_r)$
$\mathbf{V_s = AV_r + BI_r}$
$\mathbf{I_s = CV_r + DI_r}$
For a short line, A=1, **B=Z,** C=0, D=1
Line Efficiency (pu) = $V_r I \cos\phi_r / V_s I \cos\phi_s$
Voltage regulation (pu) = $(V_s - V_r)/V_r$
$= (V_{rNL} - V_{rFL}) / V_{rFL}$
$= [I(R \cos\phi_r - (or+) X\sin\phi_r)] / V_{rFL}$

4.1.8 Equivalent circuit for medium length line

A T or a π network is formed depending upon how the series impedance or the shunt admittance is lumped at a few points.

The ABCD parameters of the nominal-T network are:

$\mathbf{A} = 1+\mathbf{ZY}$
$\mathbf{B} = \mathbf{Z}(1+\mathbf{ZY}/4)$
$\mathbf{C} = \mathbf{Y}$
$\mathbf{D} = \mathbf{A}$

The ABCD parameters of the nominal-π network are:

$\mathbf{A} = 1+\mathbf{ZY}/2$
$\mathbf{B} = \mathbf{Z}$
$\mathbf{C} = \mathbf{Y}(1+\mathbf{ZY}/2)$
$\mathbf{D} = \mathbf{A}$

Nominal -T and Nominal-**π** networks are not equivalent electrically, as may be verified by using the Y-Δ transformation.

Voltage regulation (pu) = $((V_s/A) - V_{rFL}) / V_{rFL}$

4.1.9 Long line equations (above 240 km)

The solution of the voltage wave equation using the initial conditions is

$\mathbf{V} = (\cosh\gamma x)\,\mathbf{V_r} + (\mathbf{Z_0} \sinh\gamma x)\,\mathbf{I_r}$
$\mathrm{I} = (Y_0 \sinh\gamma x)\,\mathbf{V_r} + (\cosh\gamma x)\,\mathbf{I_r}$
$\gamma = \sqrt{yz} = \alpha + j\beta$
α = attenuation constant pu length
β = phase-shift constant pu length
y = shunt admittance pu length
z = series impedance pu length

$\mathbf{Z_0}$ = surge impedance = $\sqrt{(z/y)}$; $Y_0 = \mathbf{1/Z_0}$
$\mathbf{Vs} = \mathbf{AV_r + BI_r}$
$\mathbf{Is} = \mathbf{CV_r + DI_r}$
where
$\mathrm{A} = \cosh\gamma l$
$\mathbf{B} = \mathbf{Z_0}\sinh\gamma l$
$\mathbf{C} = (\mathbf{1/Z_0}) \sinh\gamma l$
$\mathbf{D} = \mathbf{A}$
l = line length

4.1.10 Equivalent circuit for a long Line

The exact equivalent π – circuit for a long line are obtained from

$\mathbf{Z\pi = B = Z_0}\mathrm{Sinh}\gamma\, l = \mathbf{(Z}\ \mathrm{Sinh}\gamma\, l)/\gamma\, l$

$\mathbf{Y\pi/2 = (A\text{-}1)/B = (}\mathrm{Cosh}\gamma\, l - 1)/\ \mathbf{Z_0}\ \mathrm{Sinh}\gamma\, l = (\tan(\gamma\, l/2).Y/2)/(\gamma\, l/2).$

The elements of the exact equivalent T – circuit are obtained from

$\mathbf{Z_T/2 = (A\text{-}1)/C = (}\mathrm{Cosh}\gamma\, l - 1)/\ \mathbf{((1/Z_0)}\ \mathrm{Sinh}\,\gamma\, l)$

$\mathbf{Z_T/2 = Z_0\ \tanh}(\gamma\, l/2) = \mathbf{(Z\ \tanh}(\gamma\, l/2))/(\gamma\, l/2)$

$\mathbf{Y_T = C = (1/Z_0)}\ \mathrm{Sinh}\,\gamma\, l = \mathbf{(Y}\ \mathrm{Sinh}\,\gamma\, l)/\gamma\, l$

4.1.11 Surge impedance loading of lines

Incident and reflected voltages on long lines

$\mathbf{V_s} = (1/2)\ (\mathbf{V_r + I_r Z_0})\ e^{d\, l}\, e^{j\beta l} + ((1/2)\ (\mathbf{V_r - I_r Z_0})\ e^{-\alpha l}\, e^{-j\beta l}$

$\mathbf{I_s} = (1/2)\ (\mathbf{V_r Y_0 + I_r})\ e^{d\, l}\, e^{j\beta l} + ((1/2)\ (\mathbf{V_r Y_0 - I_r})\ e^{-\alpha l}\, e^{-j\beta l}$

The first and second terms in each of the above equations refer to the incident and reflected voltages respectively.

The wavelength is defined by,

$\lambda = 2\pi/\beta$

The velocity of propagation ν of the waves is given by,

$\nu = \lambda f$

$\lambda = 6000$ km at 50 Hz.

When the line is terminated in its surge impedance $Z_r = Z_o$, there is no reflected wave. (Infinite line)

Surge Impedance Loading (SIL) of a transmission line

SIL = $[V_r\ (L - L)\ (\text{in kV})]^2/Z_o'$ (MW)

where $Zo' = \sqrt{\dfrac{L}{C}}$

SIL is a measure of the maximum power that can be delivered over a line. The following factors affect the maximum power:

- Line length
- Terminal apparatus impedances
- All other factors that affect stability.

To increase SIL, kV_r can be increased and Z_o reduced by using series compensation.

The distinction between maximum power and SIL should be mentioned.

4.1.12 Ferranti effect

The parameter $\mathbf{A} = \mathrm{Cosh}\,\gamma\, l$ decreases with increase in line length. In such cases V_r is considerably greater than V_s, when the line is charged but unloaded. In underground cables, the effect is much more pronounced, even in short lengths. It is called the *Ferranti effect*. Discuss the effects of shunt compensation and reactive loading.

4.1.13 Advantages of bundled conductors

- Reduced line reactance.
- Reduced voltage gradient.
- Increased corona critical voltage, and therefore, less corona power loss, audible noise, and radio interference.
- Reduced amplitude & duration of high frequency conductor vibration.

4.1.14 Disadvantages of bundled conductors

- Increased ice & wind loading.
- Inspection more complicated ,spacers required.

- Increased clearance requirements at structures.
- Increased charging kVA which may be a disadvantage at light loads.

D_s = GMR of subconductors

d = distance between two sub-conductors

D_s^b = GMR of bundled conductor

$D_s^b = (Dd)^{1/2}$ (For a 2–conductor bundle)

$D_s^b = (Dd^2)^{1/3}$ (For a 3–conductor bundle)

$D_s^b = (Dd^3)^{1/4}$ (For a 4–conductor bundle)

Average inductance per phase of a bundled conductor,

$L = 2 \times 10^{-7} \ln (D_{eq}/ D_s^b)$, H/m

$D_{eq} = (D_{12} D_{23} D_{31})$

D_{ij} = spacing between phase i and phase j

4.1.15 Factors affecting mechanical design of overhead lines

1. Character of line route
2. Right-of-way
3. Mechanical loading
4. Required clearances
5. Type of supporting structures
6. Conductor
7. Type of insulators
8. Joint use by other utilities

4.1.16 Factors affecting span length

1. Character of route
2. Proper clearance between conductors
3. Permissible tensions under maximum mechanical load

There are five kinds of stresses on lines & supports

1. Tensile
2. Compressive
3. Shearing
4. Bending
5. Twisting stress or torque

4.1.17 Sag and Tension Analysis of Overhead Lines

Required clearances:

The data for the following clearances of different voltage levels should be known.

1. Clearance of conductors passing by buildings
2. Minimum clearances of conductors above ground or rails
3. Crossing clearances of wires carried on different supports
4. Horizontal clearances at support between line conductors based on sags.

Sag and Tension Analysis :

Factors affecting sag are :

1. Conductor load per unit length
2. Span
3. Temperature
4. Conductor tension
5. Level at supports

Conductor load depends on :

1. Weight of conductors
2. Weight of ice or snow on conductors
3. Wind blowing against wire

Effect of change in temperature :

If the conductor stress is constant and if the temperature changes, the change in length is

$\Delta l = l_0 . \alpha . \Delta t$

$\Delta t = t_1 - t_0 =$ change in temperature

$\Delta l = l_1 - l_0 =$ change in length

$\alpha =$ Coefficient of linear expansion of conductor / °C. If temperature is constant while conductor stress changes (*i.e.* loading), the change in length is

$\Delta l = l_0 . \Delta T/MA$

$\Delta T = T_1 - T_0 =$ change in tension in kg

$M =$ modulus of elasticity of conductor

$A =$ Metal cross-section of conductor.

Consider the following in sag & tension calculations:

1. Supports at same level (*i*) Catenary method , (*ii*) Parabola method
2. Supports at different levels (unsymmetrical spans)
3. Effect of ice
4. Effect of wind

Line location

1. Profile & plan of right-of-way
2. Templates for locating structures

These are used to provide the following :

(*a*) Maintenance of proper clearance from conductor to ground and to crossing conductors.
(*b*) Economic layout.
(*c*) Proper grading of structures.
(*d*) Prevention of excessive insulator swing or uplift at structures.

4.1.18 Corona

If an alternating potential is applied to two wires whose spacing is large in comparison with the diameter and the potential difference is gradually increased, a point will be reached when a faint luminous glow of violet colour will appear, and a hissing sound will be heard. This phenomenon is known as Corona. The formation of corona is accompanied by a loss of power. It causes non-sinusoidal nature of current and interference with neighbouring communication circuits.

Corona formation takes place due to ionization of a layer of air immediately surrounding the conductor. For air under ordinary conditions near sea level & without impurities, the value of potential gradient at which ionization takes place can be taken as 30 kV/cm (peak).

Interference with communication circuits may be due to both electromagnetic and electrostatic action, the former producing currents, which are superposed on the true speech currents, thereby setting up distortion and the latter raising the potential of the communication circuit as a whole.

Disruptive Critical Voltage

The maximum potential gradient, g_r at the surface of the conductor is:

$$g_r = \frac{V}{r.In\left(\frac{d}{r}\right)}$$

For visual corona at normal temperature & pressure,

$$V = \frac{30\,(r+0.3\sqrt{r})}{In\left(\frac{d}{r}\right)} \text{ kv (peak)}$$

Conditions affecting corona:

- Line voltage
- Ratio d/r
- Contour of the surface
- State of the surface

Considering the above factors, the critical disruptive voltage to neutral becomes

$V_c = m_0 g_0\, \delta\, r \ln (d/r)$

m_0 = irregularity factor

g_0 = disruptive critical voltage gradient for air in kV at NTP (21.1 kV/cm, RMS)

δ = air density factor = 392 b/(273+t)

b = atmospheric pressure in cm of Hg

t = temperature in °C.

The visual critical voltage is given by

$$V_v = m_o g_o \delta r \left(1+\frac{0.3}{\sqrt{\delta r}}\right) \ln\left(\frac{d}{r}\right)$$

Power loss due to corona

Corona formation results in power loss. Peek's formula for corona loss is:

$$P = 241\left(\frac{f+25}{\delta}\right)\sqrt{\frac{r}{d}}\left(V_{ph} - V_c\right)^2 \times 10^{-5} \text{ kW/ph}$$

where V_{ph} and V_c are the effective phase and critical disruptive voltages , f is the frequency of the system, l = length in km.

Peterson's formula for corona loss is :

$$P = \frac{0.000021\, fV^2 F}{\left[\log_{10}\left(\frac{d}{r}\right)\right]^2}$$

P = power loss in kW per km of conductor under fair weather conditions.

f = frequency, Hz

V = line to ground voltage

D = spacing between conductors

R = radius of the conductor

F = corona factor determined by test

4.1.19 Audio Noise

When corona is present on the conductors, EHV lines generate audible noise, which is especially high during polluted weather. The noise is broadband , which extends from very low frequency to about 20 kHz. Corona discharges generate positive & negative ions, which are alternately attracted & repelled by the periodic reversal of polarity of the a.c. excitations. Their movement gives rise to sound-pressure waves at frequencies of twice the power frequency and its multiples, in addition to the broadband spectrum which is the result of random motions of the ions. Audible noise can become a serious problem from 'psychoacoustics' point of view, leading to insanity due to loss of sleep at night to inhabitants residing close to an EHV line.

4.1.20 Radio Interference (RI)

Pulse type corona discharge from transmission line conductors gives rise to interference to radio broadcast in the range of 0.5 MHz to1.6 MHz.

Electromagnetic Effect

The e.m.f. induced in the communication circuit due to neighbouring power circuit depends on its distance with respect to the power line. The net e.m.f. induced due to electromagnetic coupling with a 3-phase line is small since the phasor sum of induced e.m.fs. tends to zero. However, the presence of certain harmonics would cause seriously high induced e.m.fs. This problem is more serious these days since the power line current is not sinusoidal because of he use of static controllers.

Electrostatic Effect

The communication line may acquire dangerously high potential due to electrostatically induced charges. The interference between power & communication lines can be reduced considerably by transposing the conductors of both power & communication lines. The communication line may require electrostatic shielding to overcome electrostatic interference.

4.1.21 Insulators for overhead lines

Materials & types of insulators

The insulators used in connection with overhead systems employing bare conductors are composed almost invariably of glazed porcelain. Glass has also been used for medium voltages. The porcelain used should be ivory white, sound, free from defects and thoroughly vitrified .

There are three types of insulators for overhead lines:

1. Pin-type
2. Suspension type
3. Strain type
   (a) What is the difference between "puncture voltage" and "spark-over voltage "?
   (b) What the difference between arcing distances is under "wet""and" 'dry' conditions?
   (c) What is the "tracking distance ?
   (d) Tabulate the ratio of spark-over voltage to working voltage for different voltage levels.
   (e) What are the merits of suspension insulator string ?

- Each insulator is designed for a comparatively low working voltage, usually about 11 kV, and the insulation for any required system voltage can be obtained by using a "string" of such insulators.
- In the event of failure of an insulator, one unit, instead of the whole string, has to be replaced.
- The mechanical stresses are reduced.
- In the event of an increase in the operating voltage of the line , this can be met by adding the requisite number of units in each string.

What is the difference between suspension & strain insulators ?

Potential distribution over a string of insulators

1. Draw the equivalent circuit of string of three insulators.
2. Show how you would determine the potential distribution across the above string.

Model questions

1. " An insulator for overhead line should be designed so that it will spark-over before it will puncture". Why ?
2. Why is wet S.O.V less than dry S.O.V ?
3. What is the effect of pollution on S.O.V ?

String efficiency

= S.O.V of a string of n insulators / (n × S.O.V of one insulator)

The string efficiency depends on the ratio= capacitance per insulator /capacitance to earth.

Methods of improving string efficiency

The string efficiency can be improved by the following methods :

- By increasing the ratio
  m = insulator self-capacitance / capacitance to earth
  This would require long cross-arms and hence is not economical.
- Grading of the units.

This approach requires units of different sizes. Hence, it is not generally preferred. The self-capacitance of the lowest unit has to be maximum and as we move upward , the self-capacitance should decrease progressively.

- Static shielding

The voltage distribution is controlled in this method by the employment of a grading or guard ring, which usually takes the form of a large metal ring surrounding the bottom unit and connected to the metal work at the bottom of this unit, and therefore to the line. This ring, or shield , has the effect of increasing the capacitances between the metal work and the line.

The string efficiency increases with the guard ring.

Here special features of the transformer bushing may be explained.

What is the effect of surface leakage resistance on the potential distribution across a string of insulators ?

What is the effect of corona on string efficiency ?

4.1.22 Distribution System Planning

An Excel spreadsheet module can demonstrate the basics of distribution system planning. The student is to select the proper conductors and the number of shunt capacitors for compensation subject to the requirements on voltage regulation, losses and fixed and operating costs .The customer demands are specified either in power or in impedance .The operating costs for losses are specified. The capital costs for various conductor line building and for capacitor placement are specified. The student is to select the best combination of conductors and capacitors to minimise cost over a certain time frame, normally one year.

4.1.23 Tests on Electrical Materials

Type Tests – Tests carried out to prove conformity with the specifications. These are intended to prove the general qualities and design of a given type of manufactured item.

Routine Tests-Tests carried out on each part / item manufactured to check parameters (as per requirements, which are likely to vary during production.

Acceptance Tests- Tests carried out on samples taken at random from offered lot of manufactured item for the purpose of acceptance of lot.

Testing transmission line materials – Indian Standards

ACSR Conductors for 400 kV and above IS: 398 (Part 5) –1992

TYPE TESTS

No.	Type Test	Purpose
1.	Visual examination	To verify good workmanship and surface finishing of the conductor
2.	Measurement of diameter of individual aluminium & steel wires	To measure actual diameter of each strand to check that it is within specified limits
3.	Measurement of lay ratio of each layer	To measure actual lay ratio of each layer strand to check that it is within specified limits
4.	Breaking load test (on complete conductor)	To measure actual breaking load of complete conductor to check that it is within specified limits

5. Ductility test(for galvanized steel wires only). Torsion & elongation test	To record fracture of strand- number of complete twist shall not be < 18/16 for sample before /after stranding respectively Elongation shall not be < 3.5%
6. Wrapping test	For no-break observation in aluminium/galvanized steel wire (strand) after wrap/unwrap process
7. Resistance test	To check resistance of aluminium strand at 20°C
8. Galvanizing test (for galvanized steel wires only)	To check uniformity of zinc coating (4 dips of 1 min. each in $CuSO_4$ solution of sp. Gravity 1.186). The weight of zinc coating shall not be less than specified value.
9. Surface condition test	To verify cylindrical shape and relative of 50% of ultimate breaking load of the nominal movement of strands under tension condition aluminium area 100 sq. mm and conductor. (Applicable to conductors of above)
10. Corona test	To check corona extinction voltage not less than specified value.
11. Radio interference voltage test	To check RI voltage level within limits.

ROUTINE TESTS: shall be same as Acceptance Tests and shall be carried out on each coil.

ACCEPTANCE TESTS: same as Type test Nos. 1-8 given above.

SAMPLING Criteria:

- Manufacturer shall normally take samples of individual wires for the acceptance tests 1-8 above before stranding of not less than 10% of wire coils.
- Alternatively, if desired by the purchaser at the time of placing an order that the tests be carried out in presence of the purchaser's representative. Samples shall then be obtained by cutting 1.2 meters from outer end of finished conductor from not more than 10% of the drums.
- Coils offered for inspection shall be divided into two equal lots , the number of lots being equal to the number of samples to be selected, a fraction of a lot being counted as a complete lot. One sample coil shall be selected at random from each lot.

  Earth Wires (Galvanized Strands for Earthing) IS: 12776-1989

The tests under TYPE, ROUTINE, and ACCEPTANCE category are not specified in the Indian Standard. However, the following tests shall be carried out on the selected samples.

No.	Test	Purpose
1	Breaking load test	To verify strength/measure ultimate breaking load of galvanized wires separately to be within limits
2	Elongation test	For elongation to be within limits
3	D.C. Resistance test	Actual resistance of wire to be within limits
4	Wrapping test	To verify capacity to withstand wire twisting. No break in wire after wrap/unwrap process
5	Galvanizing test	To check uniformity of zinc coating. No permanent copper deposition after 4 dips of 1 min. each in $CuSO_4$ solution of sp. gravity 1.186. The weight of zinc coating shall not be less than specified value.
6	Torsion test	To verify capacity to withstand torsion of wire. No break in wire after process.

REJECTION & RE-TEST : If test sample fails any of the tests, three further samples from same lot, out of which, one sample from same drum of original sample, is selected and the tests repeated on all three samples.

Porcelain Disc Insulators for Overhead Lines with Nominal Voltage > 1000 V, IS 731-1989

TYPE TESTS

NO. Type test	Purpose
1. Visual examination	For workmanship /surface defects
2. Verification of dimensions	For ensuring dimensions as per requirement and approved drawing
3. Visible discharge test	For measuring visual corona
4. Impulse voltage withstand test	To check ability of the insulator housing towithstand voltage stresses under dry and wet conditions
5. Wet power frequency voltage withstand test	To check ability of the insulator housing to withstand voltage stresses under wet conditions
6. Temperature cycle test	To check capability of the insulators to withstand thermal stresses
7. Electro-mechanical failing load test	To check capability of the insulators to mechanical withstand under combined electrical and –Type B stresses (For string insulator units only)
8. Mechanical failing load test	To check capability under mechanical stresses(For string insulators of Type A and those of Type B to which electro-mechanical failing load test is not applicable, and for rigid insulators only)
9. 24 Hours mechanical strength test	To check capability to withstand the electrical stresses (for insulators- Type B only)
10. Puncture test	To check capability to withstand the electrical stresses (for insulators –Type B only)
11. Porosity test	To confirm non-porous nature of product
12. Galvanizing test	To determine the uniformity and thickness of zinc coating

Notes:

1. A radio interference test is under consideration
2. Type tests are normally carried out once and unless otherwise agreed to, test certificates giving results of type tests , made on not less than two insulators identical in all essential details with those to be supplied, are regarded as evidence of compliance. The tests should be carried out in the order mentioned below:
   (a) On both insulators: Tests 1-6 (b) On first insulator: Tests 9,7,8,11
   (c) On second insulator: Tests 10 &12.
3. Type tests shall be carried out and certified by the manufacturer or by an agreed independent authority.

ROUTINE TESTS

No.	*Routine test*	*Purpose*
13	Visual examination	For workmanship /surface defects
14	Mechanical routine tests	To confirm withstanding mechanical stresses during normal conditions (for string insulator units only)
15	Electrical routine tests	To confirm withstanding electrical stresses during normal conditions (for Type B string insulators and rigid insulators)

ACCEPTANCE TESTS: Test Nos. 5, 6, 9, 7, 8, 10, 11, 12

SAMPLING CRITERIA: IS: 731, 1987

FOR DIMENSIONS & TEMPERATURE CYCLE TESTS

Lot size, N	*First sample size, n1*	*Second sample size, n2*	*Permissible failure, a*	*First rejection number, r1*	*Second rejection number, r2*
Up to 1000	8	8	0	2	2
1001- 3000	13	13	0	2	2
3001-10000	20	20	0	2	2
10001 & above	32	32	1	3	4

FOR MECHANICAL, ELECTRO-MECHANICAL and POROSITY TESTS

Lot size, N	FOR MECHANICAL, ELECTRO-MECHANICAL and POROSITY TESTS		FOR GALVANIZING & PUNCTURE TESTS	
	First sample size, n1	*Second sample size, n2*	*First sample size, n1*	*Second sample size, n2*
Up to 1000	5	5	3	3
1001- 3000	8	8	5	5
3001-10000	13	13	7	7
10001 & above	20	20	12	12

NOTE: The samples selected shall be divided approximately into three parts and subjected to the applicable tests in the following order.

Parts of sample	Tests on string insulator units Type A	Tests on string insulator units Type B	Tests on Rigid insulators Type A	Tests on Rigid insulators Type B
First & second part	5, 6, 8,11 & 9	5, 6,7 or 8 (whichever is applicable), 9 (when specified) & 11	5, 6, 8, & 11	5, 6, 8, & 11
Third part	5, 6, 10 & 12	5, 6, 10 & 12	5, 6, 10 & 12	5, 6, 10 & 12

Other relevant Indian Standards are the following:

- AAA Conductors [All Aluminium Alloy Stranded Conductors] IS: 398 (Part 4) –1994
- ACSR Conductors IS: 398 (Part 2) –1994
- Pre-formed armour rods for conductor, IS: 2121 (Part I) –1981
- Repair Sleeve and Mid-span joints for conductors, IS: 2121 (Part II) –1981
- Stockbridge Vibration Dampers for overhead Lines IS: 9708-1993
- Spacer and Spacer Dampers for Twin Horizontal Conductors IS: 10162-1982
- Earth-wire accessories- Suspension and Tension Clamps; Mid-span Compression Joints; Repair
- Sleeves and Flexible Copper Bonds IS2121 (Part-3)-1992
- Cross-by Clips IS2121 (Part-3)-1992
- H Frame/Tower Structural Steel [rail, Girder, Angle etc.] IS: 2062
- Stay wires [Hot-dipped galvanized Stay Brand IS: 2141-1992
- Porcelain Insulators-Bus Post for system nominal voltage > 1000 V IS: [2544-1973]
- Step Bolts and Their Nuts for steel structures –IS: 10238 –1982

OBJECTIVE TYPE QUESTIONS

1. The surge impedance of a 110 kV, 3-phase transmission line is 440 Ω. The surge impedance loading of the line is

(a) $\sqrt{3}\,(110)^2/440$ MW (b) $(110)^2/440$ MW

(c) $(110)^2/\sqrt{3}\,110)^2$ MW

Ans. (b)

2. The capacitance and inductance per unit length of a 3-phase line, operating at 110 kV are 0.01 μF and 2.5 mH. The surge impedance of the line is

(a) 50 Ω (b) 500 Ω (c) 250 Ω

Ans. (b)

3. A long transmission line is energized at the sending end and is kept open circuited at the receiving end. The magnitudes of the sending end voltage V_s and of the receiving end voltage Vr satisfy the following relationship

(a) $V_s = V_r$ (b) Vs is greater than Vr (c) Vs is less than Vr

Ans. (c)

4. Voltage regulation of a short transmission line is

(a) always positive (b) always negative

(c) either positive, negative, or zero

Ans. (c)

5. The capacitance of an overhead line increases with

(i) increase in mutual geometric mean distance

(ii)increase in height of conductors above ground

(a) Both are true (b) Both are false

(c) Only (i) is correct

Ans. (b)

6. Shunt compensation for long EHV lines is primarily resorted to

(a) improve voltage profile (b) improve stability

(c) reduce fault currents

Ans. (a)

7. Series compensation is primarily resorted to

(a) improve voltage profile (b) improve stability

(c) reduce fault currents

Ans. (b)

8. Fair weather corona loss may be computed using the empirical formula given by Peterson. Accordingto Peterson's formula corona loss is proportional to

(a) f and V^2 (b) f^2 and V

(c) f and V

where *f* and V are the system frequency and voltage respectively.

Ans: (a)

9. Bundled conductors are used in EHV lines primarily for

(a) reducing cost of the line

(b.) reducing corona loss and radio interference

(c) increasing stability limit.

Ans. (b)

10. There are 20 discs in the string of insulators of a 3-phase 400 kV transmission line. String efficiency is 80 %. The maximum voltage across any disc is

(a) 25 kV (b) $25/\sqrt{3}$ kV

(c) $25\sqrt{3}$ kV

Ans. (b)

11. Two or three sheds or petticoats are provided in pin-type insulators in order to increase

(a) creepage resistance (b) spark-over voltage (S.O.V)

(c) puncture voltage

Ans. (a)

12. Pin -type insulators are used up to

(a) 11 kV (b) 33kV

(c) 132kV

Ans. (b)

13. Insulators used for transmission line at the dead -end tower are

(a) suspension type (b) shackle type

(c) strain type

Ans. (c)

14. Economic studies have shown that D.C. transmission is cheaper than a. c transmission for lengths
(a) below 300 km (b) beyond 600 km
(c) beyond 1200 km
Ans. (b)

15. Transmission voltages in the range 230 kV-765 kV are known as
(a) High voltage (b) Extra High Voltage
(c) Ultra High Voltage
Ans. (b)

16. Which one of the following statements is false? As the transmission voltage increases,
(a) Corona loss decreases
(b) Conductor copper loss decreases
(c) Cost of insulators, transformers, switches & circuit breakers increases
Ans. (c)

17. The internal inductance of a solid conductor of radius r and carrying a current I is equal to
(b) $0.5\,I \times 10^{-7}$ H/m (b) $0.5\,I \times 10^{-7}\exp(-1/4) \times r$ H/m
(b) 0.5×10^{-7} H/m
Ans. (c)

18. Which one of the following statements is *not* true?
(a) The GMD method of finding inductance does not apply to ACSR conductors
(c) Current density in ACSR conductors is uniform
(c) The GMD between two circular areas, each of different diameters, is equal to the distance between their centres.
Ans. (b)

19. Expanded ACSR conductors are used
(a) To increase the tensile strength of the line
(b) To reduce corona loss
(c) To reduce I^2R loss
Ans. (b)

20. A conductor with 19 strands, each of same diameter and each having an inductance of L Henries is used for a transmission line. The total inductance of the conductor will be
(a) L/19 (b) L/361
(c) 19L
Ans. (a)

21. The line to neutral capacitance of single -phase line with conductors of radius 1cm and spaced 1m apart is equal to
(a) $10^{-9}/72$ F/m (b) $10^{-9}/36\pi$F/m (c) $2\pi^2 10^{-7}$ F/m
Ans. (a)

22. In a double-circuit line with hexagonal spacing
(a) The phases are balanced, but the conductors of each individual phase are not balanced.
(b) The conductors of each individual phase are balanced , but the phases are not balanced
(c) The phases, and the conductors of each individual phase are both balanced
Ans. (c)

23. Which one of the following statements is true ?
(a) Skin effect at 50 Hz is negligible for larger diameter conductors but becomes appreciable for smaller conductor.
(b) Skin effect at 50 Hz is negligible whatever the diameter of the conductor.
(c) Skin effect at 50 Hz is negligible for the smaller diameter conductors but becomes appreciable for the larger conductor.
Ans. (c)

24. Which one of the following statements is true ?

(a) Resistance of a conductor decreases and the internal inductance increases as the frequency is increased.

(b) Resistance and internal inductance of a conductor both increase with increase of frequency.

(c) Resistance of a conductor increases and the internal inductance decreases as the frequency is increased.

Ans. (c)

25. The surge impedance of a double-circuit power transmission line is

(a) 40 Ω (b) 200 Ω

(c) 400 Ω (d) 800 Ω

Ans. b

26. The surge impedance of a telephone line is

(a) 50 Ω (b) 75 Ω

(c) 200 Ω (d) 400 Ω

Ans. (b)

26. Bundle conductors are preferred in EHV transmission lines because

(a) It is easy to fabricate thin conductors and combine them to make a bundle

(b) Inductance of the line is reduced, and the corona loss, and radio & TV interference is minimized.

(c) Tower height is reduced and hence transmission cost is low.

Ans. (b)

27. Inductive interference between power & communication lines can be minimized by

(a) Increasing the spacing of power line conductors

(b) Transposing power line conductors

(c) Transposing communication line conductors

(d) Either (b) or (c).

Ans. (d)

28. The percentage regulation of an overhead transmission line can be zero when the load power factor is

(a) Lagging

(b) Unity

(c) Leading.

Ans. (c)

29. Which one of the following statements is true ?

(a) Skin effect increases the resistance of a conductor ,but proximity effect decreases the resistance.

(b) Both skin effect and proximity effect increase the resistance of a conductor.

(c) Both skin effect and proximity effect increase the internal inductance of a conductor.

Ans. (b)

30. A transmission line having parameters A1, B1, C1, D1 is in parallel with another having parameters A2, B2, C2, D2. The overall " A " parameter of the combination is

(a) A1A2 + B1C2

(b) (A1B2 + A2B1)/(B1 + B2)

(c) C1+C2 + (A1-A2)(D2-D1)/(B1+B2)

Ans. (b)

31. Disruptive critical voltage is

(a) Equal to (b) Greater than

(c) Less than visual critical voltage for corona on an overhead line.

Ans. (c)

32. Corona loss

(a) Increases

(b) Decreases

(c) Does not change the switching voltage on a transmission line.

Ans. (b)

33. Handling of the telephone receiver may become dangerous due to

(a) Electromagnetic induction

(b) Electrostatic induction

(c) Both electromagnetic and electrostatic induction from a power line.

Ans. (b)

34. Mutual inductance between a three - phase power line and a telephone line due to third harmonic current in the power line is

(a) The algebraic sum of mutual inductances from individual phase wires

(b) The arithmetic sum of mutual inductances from individual phase wires

(c) Zero

Ans. (b)

35. Dielectric strength of mechanically sound porcelain is

(a) 10 kV/cm

(b) 22 kV/cm

(c) 65 kV/cm

(d) 100 kV/cm

Ans. (c)

36. Dielectric strength of glass is

(a) 22 kV/cm

(b) 60-66 kV/cm

(c) 140 kV/cm

(d) 240 kV/cm

Ans. (c)

37. Which one of the following statements is true ?

(a) For *11 kV* insulators, the ratio of *wet* spark-over voltage to working voltage is 8.3

(b) For *66 kV* insulators, the ratio of *dry* spark-over voltage to working voltage is 8.3

(c) For *11 kV* insulators, the ratio of *dry* spark-over voltage to working voltage is 8.3

Ans. (c)

38. Which one of the following statements is true ?

(a) Flash-over tests are performed on all insulators

(b) Routine tests are performed on 1/2 percent of insulators supplied

(c) Design tests are done on 1/2 percent of insulators supplied

Ans. none of the above

39. Which one of the following statements is true ?

(a) The spark-over voltage is less than the puncture voltage

(b) The spark-over voltage is greater than the puncture voltage

(c) The spark-over voltage is equal to the puncture voltage

Ans. (a)

40. Which one of the following statements is true ? The sheds of an insulator should be shaped

(a) To conform to the electrostatic tube of force and the body should be shaped to conform to the equipotential surfaces.

(b) To conform to the equipotential surfaces and the body should be shaped to conform to the electrostatic tubes of force.

(c) To conform to the equipotential surfaces and so also the body.

Ans. (b)

41. In a suspension insulator, the mechanical stresses on the conductor are
(a) Increased
(b) Reduced
(c) The same
Ans. (b)

42. The string efficiency of the insulator can be increased by
(a) Increasing the number of strings in the insulator
(b) Increasing the ratio , capacitance to earth / capacitance per insulator
(c) By the correct grading of various capacitances
(d) Decreasing the number of strings
Ans. (c)

43. String efficiency of insulators for wet flash-over is
(a) Less
(b) More
(c) The same as that for dry flash-over
Ans. (b)

44. The potential across insulator discs can be equalized by having
(a) The same capacitance for each unit.
(b) The highest capacitance for the lowest unit and decreasing progressively the capacitance of other units.
(c) The lowest capacitance for the lowest unit and increasing progressively the capacitance of other units.
Ans. (b)

45. Grading ring serves the purpose of
(a) Equalizing the voltage distribution across discs.
(b) An arcing shield
(c) Both equalizing the voltage distribution and acting as an arcing shield
Ans. (c)

PROBLEMS WITH SOLUTIONS

Prob.4.2.1 Nominal-T-ABCD constants

Determine the ABCD constants for the nominal T circuit of a transmission line for which R = 10 Ω, X= 20 Ω, and Y = 400 μ Siemens for each phase.

Solution: R = 10 Ω

X = 20 Ω

Y = 0.0004 S

Z = R + jX

N.B. Bold letters indicate complex numbers

YZ = – jYR + YX

YR = 0.004

YX = 0.008

A = 1+ (**YZ**/2)

ReA=1+(YX)/2 = 0.996

ImA = –YR/2 = – 0.002

A = 1 + (**YZ**/2) = 0.996 – j.002 **Answer**

B =**Z** (1 + **YZ**/2) = Z (**ReA** + jImA) =(R + jX).(ReA + jImA)

ReB = R × ReA–X × ImA = 10

ImB = X × ReA + R × ImA = 19.9

B = 9.94 + j20.08 Ω **Answer**

C = **Y** = j.0004S **Answer**

Prob.4.2.2 Overhead line-Basic Parameters and surge impedance

A single-circuit three-phase 60 Hz transmission line consists of three conductors 1, 2, and 3. each of diameter 0.635 cm. The spacings between the conductors are $D_{12} = D_{13} = 5$ m and $D_{23} = 8$ m. Find the inductive and capacitive reactance of the line per km per phase. Determine also the surge impedance of the line and the phase shift in rad/m.

Solution. f = 60 Hz

$\mu_0 = 4 \times 3.14 \times 1E - 07 = 0.00000126$

$D_{12} = 5$ m

$D_{13} = 5$ m

$D_{23} = 8$ m

$D_{eq} = \sqrt{D_{12}.D_{23}.D_{31}} = 5.848$ m

d = 0.00635 m

r = d/2 0.00318 m

$D_{eq} = 5.848$

$X_L/m = f\mu_0 \ln (D_{eq}/r) = 0.000567$ Ω/m = 0.567 Ω/km

L /m =1.503E– 06 H/m

$\varepsilon_0 = 8.846E - 12$ F/m

C per m = $2\pi\varepsilon_0 / \ln (D_{eq}/r) = 7.389E{-}12$ F/m

$X_c/m = 1/ (2 \times 3.14 \times f \times C) = 359166276$ Ω /m = 359166.3 Ω/km **Answer**

Z_o surge impedance $\sqrt{\frac{L}{C}} = \sqrt{X_L \times X_C} = 451.113$ Ω **Answer**

$\omega = 2\pi f$

β, phase shift, rad/m = $\omega\sqrt{LC}$ =1.256E-06 rad/m **Answer**

Prob.4.2.3 Overhead lines - Potential distribution across a string of insulators.

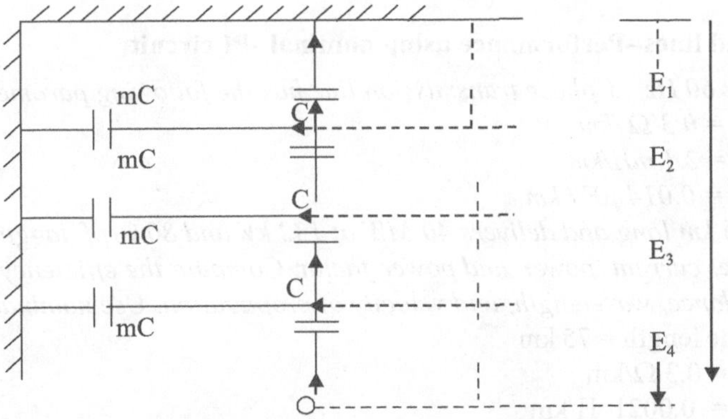

Fig.4.2.3. *Equivalent circuit of insulator string*

Fig. 4.2.3 shows a string of suspension insulators m = 5. The voltage between the line conductor and earth is 76 kV. Find

(i) *the potential difference across each unit, and*

(ii) *the string efficiency.*

Solution. Writing KCL equations at nodes 1, 2 and 3:

$mCE_2 = mCE_1 + E_1$ . . . (1)

$mCE_3 = CE_1 + CE_2 + mCE_2$ . . . (2)

$mCE_4 = CE_1 + CE_2 + CE3 + mCE_3$ . . . (3)

Also;

$E = E_1 + E_2 + E_3 + E_4$ . . . (4)

m = 5

E = 76 kV

Rewriting eqs.1–3 :

$$E_2 = E_1(1+\frac{1}{m}) \quad \ldots (5)$$

$$E_3 = E_1\left(1+\frac{3}{m}\right)+\left(\frac{1}{m^2}\right) \quad \ldots (6)$$

$$E_4 = E_1\left(\frac{1+6}{m}\right)+\left(\frac{5}{m^2}\right)+\left(\frac{1}{m^3}\right) \quad \ldots (7)$$

From equation (4)

$$E = E_1\left(4+\frac{10}{m}\right)+\left(\frac{6}{m^2}\right)+\left(\frac{1}{m^3}\right)$$

or,

$$E_1 = \frac{E}{4+\left(\frac{10}{m}\right)+\left(\frac{6}{m^2}\right)+\left(\frac{1}{m^3}\right)}$$

E_1= 12.164 kV **Answer**

Using equation (5) to (7)

$$E_2 = E_1\left(1+\frac{1}{m}\right) = 14.597\text{kV}$$ **Answer**

$$E_3 = E_1\left(1+\left(\frac{3}{m}\right)+\frac{1}{m^2}\right) = 19.949 \text{ kV}$$ **Answer**

$$E_4 = E_1\left(1+\left(\frac{6}{n}\right)+\frac{5}{m^2}+\frac{1}{m^3}\right) = 29.291 \text{ kV}$$ **Answer**

Prob.4.2.4 Overhead lines -Performance using nominal -PI circuit

A single-circuit 60 Hz, 3-phase transmission line has the following parameters:

R = 0.3 Ω/km

L = 2.1 mH/km

C = 0.014 μF / km

If this line is 75 km long and delivers 40 MW at 132 kV and 80 % pf lagging, determine the sending -end voltage, current, power, and power factor. Compute the efficiency of transmission, characteristic impedance, wavelength, and velocity of propagation. Use nominal π circuit.

Solution. line length = 75 km

R = 0.3 Ω/km

L = 0.0021 H/km

C = 0.000000014 F/km

R' = Rl = 22.5 Ω

L' = Ll = 0.1575 H

C' = Cl = 0.00000105 F

f, Frequency = 60 Hz

ω = 2πf = 376.8 rad/s

X_L' = ωL' = 59.346 Ω

ωC' = 0.000396 S

Re **Z** = 22.5 Ω

Im **Z** = 59.346 Ω

Im **Y** = 0.000396 S

P Power = 40000.000 kW

Vr L-L= 132 kV

p.f. power factor = 0.8 lag

$$I_r = \frac{P}{\sqrt{3}V_r \times pf} = 218.693$$

Re **Ir** = Ir × pf =174.954 A

Im **Ir** = Ir × sin (cos^{-1}(pf)) =131.215 A

For nominal π:

$$V_s = \left(\frac{ZY}{2}+1\right)Vr.+Z.Ir$$

Vr phase value = 76210.235 Volts

Re **Vs** =Vr – (Vr × ImZ × ImY/2) + ReZ × ReIr+ImZ × ImIr = 87039.162 Volt

Im **Vs** = (Vr × ReZ × ImY/2) + ImZ × ReIr-ReZ × ImIr =7769.705 Volts

Vs phase value = 87385.262

Vs (L – L) =151355.644 Volts **Answer**

$$\theta = \text{angle Vs} = \frac{180}{\pi}\tan^{-1}\left(\frac{\text{Im}\,V_s}{\text{Re}\,V_s}\right) = 5.101°$$

$\mathbf{I_s} = (\mathbf{V_s Y}/2) + (\mathbf{V_r . Y}/2) + \mathbf{I_r}$

$$\text{Re}\,\mathbf{I_s} = \text{Im}\,Vs.\text{Im}\left(\frac{Y}{2}\right) + \text{Re}\,Ir = 173.417\text{ A}$$

$$\text{Im}\,\mathbf{I_s} = \text{Re}\,Vs.\text{Im}\left(\frac{Y}{2}\right) + Vr.\text{Im}\left(\frac{Y}{2}\right) - \text{Im}\,Ir = -98.921\text{ A}$$

I_s =199.647 A

ϕ = angle (I_s) = – 29.701°.

$$\text{p.f. power factor} = \cos\left[(-\phi \times \frac{\pi}{180°}) + (\theta \times \frac{\pi}{180})\right] = 0.821$$

sending end power = $\sqrt{3}$. $V_s.I_s.pf$ = 42985809 W= 42986 kW **Answer**

$$\text{Efficiency} = \frac{P}{P_s} \times 100 = 93.053\%$$ **Answer**

γ, propagation constant = $\sqrt{ZY}$ = a + j b = A cos (B) + j A sin (B)

where

$$A = \sqrt{(\text{Re}\,Z.\text{Im}\,Y)^2 + (\text{Im}\,Z.\text{Im}\,Y)^2} = 0.158$$

B = (1/2) × tan^{-1} ((– ReZ/ImZ)) = – 0.181 rad

a =Reγ = Acos (B) = 0.155

b = Imγ = Asin (B) = – 0.0285 rad

$$\lambda = \frac{2\pi l}{\beta} = 3023.311\text{ km}$$ **Answer**

velocity = f.λ =181398.717km/s **Answer**

$$Z_o = \sqrt{\frac{Z}{Y}} = A'\cos(B') + jA'\sin(B')$$

where

$$A = \left[\frac{(\mathrm{Re}\,Z)^2 + (\mathrm{Im}\,Z)^2}{(\mathrm{Im}\,Y)^2}\right]^{\frac{1}{2}} = 400.523$$

$$B = \frac{1}{2}\tan^{-1}\left(\frac{-\mathrm{Re}\,Z}{\mathrm{Im}}\right) \quad 0.181$$

$\mathrm{Re}\ Z_o = 393.966$

$\mathrm{Im}\ Z_o = -72.176$

$Z_o = 400.523\ \Omega$ **Answer**

angle of $Z_o = -10.386°$

Prob.4.2.5 ABCD parameters of cascaded networks

Find the ABCD parameters of a network resulting when a resistance of 10 ohms is connected in series at the sending end of the network whose parameters are as follows:

$A1 = 1\angle 0°$; $B1 = 40\angle 90°\ \Omega$; $C1 = 0.003\angle 90°$, *Siemens*, $D1 = 0.9\angle 0°$

Solution: The resistance of 10 ohms has the following parameters:

$A2 = D2 = 1$, $B2 = 10\ \Omega$; $C2 = 0$

The overall parameters of the cascaded networks are

$A_o = A_1A_2 + B_2C_1 = 1 + j400$ **Answer**

$B_o = A_2B_1 + B_2D_1 = 9 + j40\ \Omega$ **Answer**

$C_o = A_1C_2 + C_1D_2 = j.003$ Siemens **Answer**

$D_0 = B_1C_2 + D_1D_2 = 0.9$ **Answer**

Prob.4.2.6 Overhead lines- Capacitance and Charging current

A 3-phase transmission line 30km long has its conductors of 0.5 cm diameter spaced at the ends of an equilateral triangle of 120 cm side. The line is fed from a star-connected transformer (with neutral point earthed) at 33 kV (L-L) and 50 Hz. Find the charging current.

Solution. $\varepsilon_o = 8.8419\text{E-}12$ F/m

D = spacing 1.2 m

R = radius 0.0025 m

C, Capacitance $= 2\pi\varepsilon_0 \ln\left(\frac{d}{r}\right) = 8.998\text{E-}12$ F/m

l Length = 30 km

C Total = 2.699 E – 07 F

f = 50 Hz

V / phase = 1905 .567 Volts

I, charging current = V.2πfC

= 1.615 A **Answer**

Prob.4.2.7 Overhead lines -Performance using short-line representation-Example 1.

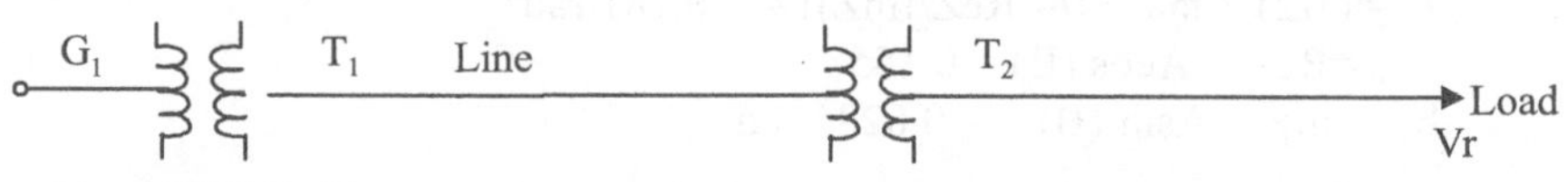

Fig.4.2.7. (*a*) *System diagram*

A load of 50 MW at 0.8 p.f. lagging exists. The internal no-load voltage of the generator is held at 17 kV (line). Determine the voltage at the load busbar. The load may be represented by a constant impedance.

T_1: *50 MVA, 11/132 kV, X = 10%*

T_2: *50 MVA, 132/33 kV, X = 12%*

Line reactance =100 Ω

Solution. Base MVA chosen 100 MVA

kVbase 132 kV

$$Z_b = \frac{(kV_{basae})^2}{Base\,\text{MVA}} = 174.24\ \Omega$$

Line X = 100 Ω

Line X pu = 0.573 pu

T1:

MVA rating = 50 MVA

Xpu on rating = 0.1 pu

Xpu on common base = Base MVA × Xpu on rating/MVA rating = 0.2 pu

T2:

MVA rating 50 MVA

Xpu on rating = 0.12 pu

Xpu on common base = Base MVA × Xpu on rating / MVA rating = 0.24 pu

V_s 17 kV

kV base generator side = 11 kV

V_s pu = V_s/ kVbase = 1.545

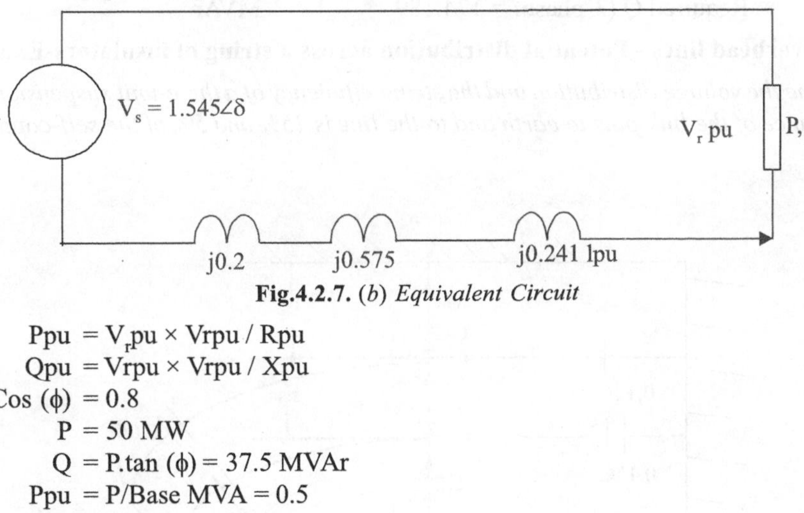

Fig.4.2.7. *(b) Equivalent Circuit*

Ppu = $V_r$pu × Vrpu / Rpu

Qpu = Vrpu × Vrpu / Xpu

Cos (ϕ) = 0.8

P = 50 MW

Q = P.tan (ϕ) = 37.5 MVAr

Ppu = P/Base MVA = 0.5

Qpu = 0.375

Rpu = 0

Xpu = 1.015pu

1/Qpu = 2.67

Xpu = 2.67 $V_r$pu

Total reactance from source to load = 1.013

$$\text{Ipu} = \frac{1.545\angle\delta}{\sqrt{R_{pu}{}^2 + (X_{pu} + 1.0139)^2}} = \frac{Vrpu}{\sqrt{R^2_{pu} + X^2_{pu}}}$$

Simplifying this expression, we get a quadratic equation:

11.13 $V_r$pu^4– 21.14 $V_r$pu^2 + 1.03 = 0.

Its solution is

$V_r$pu = 1.36 pu = 44.88 kV **Answer**

Prob.4.2.8 Short line -Required Q at the receiving end to maintain $V_s = V_r$

A 3-phase transmission line supplies 600 MW to a load of 0.98 pf lagging. The line reactance is 21.83 Ω/phase. Determine the reactive power required at the receiving end if the voltage at both sending and receiving ends is to be maintained at 275 KV (L–L).

Solution: $\mathbf{I = (Vs{-}Vr)/jX}$

$$Q_r = I_m(\mathbf{Vr \times I\ conjugate}) = \text{Im}((\mathbf{V_r V_s\ Conjugate} - V_r^2)/(-jX)$$

$$\text{Im}(V_r \angle 0°.V_s \angle\delta - V_r^2)/(-jX)$$

$$= (V_r V_s \cos\delta - V_r^2)/X$$

Vr / phase = 158.771 kV

Vs / phase = 158.771 kV

X / phase = 21.83 Ω

P / phase = 200 MW

$$\sin\delta = \frac{P_r.X}{V_s.V_r}$$

$\delta = 0.174$ rad.

$Q_r = -17.451$ MVAr

p.f. = 0.98 lag

$\phi = \cos^{-1}(\text{pf}) = 0.200$ rad.

$Q_{load} = P_r.\tan\phi = 40.611$ MVAr

Required $Q = Q_{load} - Q_r = 58.063$ MVAr/phase

Required Q (3-phase) = 174.189 MVAr **Answer**

Prob.4.2.9 Overhead lines - Potential distribution across a string of insulators-Example 2

Determine the voltage distribution and the string efficiency of a three-unit suspension insulator if the capacitance of the link-pins to earth and to the line is 15% and 5% of the self-capacitance of each unit.

Solution.

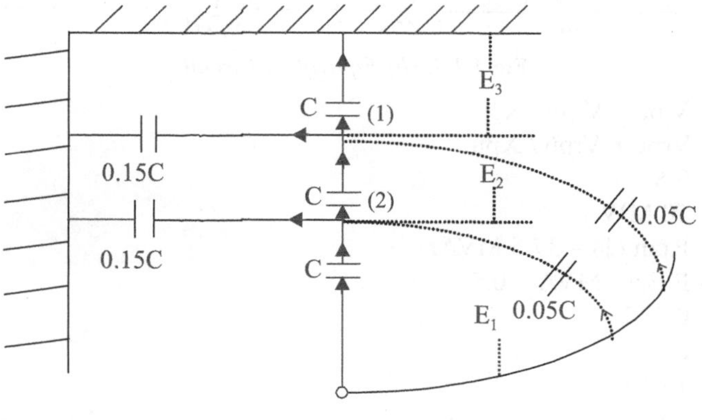

Fig. 4.2.9. *Equivalent circuit of string insulator*

Writing KCL at Node 1:

$$E_2.C = E_3.C + E_1\,0.15C - (E_1 + E_2)\,0.05C$$

Writing KCL at Node 2:

$$E_1.C = E_2.C + (E_2 + E_3)\,0.15C - E_1 \times 0.05C$$

Rewriting the two equations:

$$E_1 0.1C - E_2 1.05C + E_3.C = 0 \quad \ldots (1)$$

$$E_1 0.05C - E_2 1.15C - E_3\,0.15C = 0 \quad \ldots (2)$$

Further, let

$$E_1 + E_2 + E_3 = 100, \text{ say} \quad \ldots (3)$$

We have to solve the above equations for E_1, E_2, and E_3:

The solution is: $E_1 = 38.54$, $E_2 = 31.25$, $E_3 = 30.21$ V **Answer**

Check: $E_1 + E_2 + E_3 = 100$V

Prob.4.2.10 Overhead lines -Performance using short-line representation-Example 2

A short line has a series reactance of 16 Ω/phase. The input to the line is 20 MW/phase and 16 MVAr lagging /phase. If **Vr** *= (100 + j0) kV/phase, calculate* $\mathbf{V_s}$ *and* $\mathbf{I_s}$

Solution. $P = V_s.Vr.\sin\delta/X$

$$Q = \frac{V_s^2}{X} - V_sV_r\cos\delta / X$$

Rewrite these equations as;

$$\text{Sin}\delta = \frac{P.X}{V_s.V_r} \qquad \ldots (1)$$

$$\text{Cos}\delta = \left[\left(V_s^2/X\right)-Q\right]\left(\frac{X}{V_sV_r}\right) \qquad \ldots (2)$$

Square and add Eqns. (1) & (2)

We get the following quadratic equation in V_s^2:

$V_s^4 + bV_s^2 + c = 0$, where

$b = -2QX - V_r^2$

$c = P^2X^2 + Q^2X^2$

Here $P = 20$ MW

$Q = 16$ MVAr

$V_r = 100$ kV

$X = 16$ ohms

$b = -2QX - Vr \times Vr = -10512$

$c = P^2X^2 + Q^2X^2$

$= 167936$

$V_s^2 = 10496$ V0.0312

$V_s = 102.449$ kV

$\text{Sin}\delta = P.X/(V_s.V_r) = 0.0312$

$\text{Cos}\delta = 0.999$

Re $\mathbf{V_s} = V_s \times \cos\delta$

102.400kV/ph

Im $\mathbf{V_s} = Vs \times \sin\delta = 3.2$ kV/ph

Vs magnitude = 102.449 kV

$\mathbf{I} = (Vs\cos\delta + j\,V_s\sin\delta - V_r)/jX$

Re $\mathbf{I} = V_s \times \sin\delta / X$ 0.2 kA **Answer**

Im$\mathbf{I} = -V_s.\text{Cos}\delta + V_r - 0.150$ kA **Answer**

I mag 0.250 kA **Answer**

Prob.4.2.11 Overhead lines-Performance of parallel feeders from ABCD parameters

Two parallel feeders have the following constants:

Constants	**Feeder I**		**Feeder II**
A= D	0.816 ∠4.35°		0.871 ∠2.5°
B	227.2 ∠72.3°		∠ 73°
C	15.7×10^{-4} ∠ 91.4°	12.94×10^{-4}	∠92.35°

Calculate the sending end voltage of the system when the receiving end is on open circuit, and the receiving end voltage is 132 kV (L-L)

Solution.

$A = (A_1B_2 + A_2B_1)/(B_1 + B_2)$

A_1 mag $= 0.816$

A_1 angle $= 0.076$ rad

A_2 mag = 0.871
A_2angle= 0.044 rad
B_1 mag = 227.200
B_1 angle = 1.261 rad
B_2 mag = 196.000
B_2 angle = 1.273 rad
ReA_1 = 0.813
ImA_1 = 0.061
Re A_2 = 0.870
ImA_2 = 0.037
ReB_1 = 69.214
ImB_1 = 216.400
Re B_2 = 57.425
ImB_2 = 187.398

A = (ReA_1+ jImA_1) (ReB_2+ jImB_2) + (ReA_2+ JImA_2) (ReB_1+ jImB_1) / (ReB_1 + jImB_1) + ($R_e B_2$ + jImB_2))
= ReA_1ReB_2 – ImA_1ImB_2 + Re2ReB_1– ImA_2ImB_1+ j
(ImA_1ReB_2+ Re$A_1 I_m B_2$ + ReB1ImA_2 + ReA_2ImB_1)/
(ReB_1+ReB_2+ j (ImB_2 + ImB_1))
= N/D, say

ReN = 87.143
ImN = 346.963
ReD = 126.640
ImD = 403.799
Re **A** = (ReN.ReD + ImN.ImD) / $(ReD)^2$ $(ImD)^2$= 0.843
Im **A** = ImN.ReD-ReN.ImD) / $(ReD)^2$ $(ImD)^2$ = 0.048
A mag 0.845
Vr = 132 kV
Vs = Amag.V_r= 111.584 kV **Answer**

Prob.4.2.12 Overhead line: Earthing -Peterson coil

A 230 kV, 3-phase 50 Hz 200 km transmission line has a capacitance to earth of .02 μF/km/ph. Calculate the inductance and kVA rating of the Peterson coil used for earthing the above system.

Solution. f = 50 Hz
l = 200 km
C′ / km = 0.00000002 F
C, Total = 0.000004 F
ω = 314 rad/s
L =1/ $(3\omega^2 C)$ = 0.845 H **Answer**
Vph = 132.947 kV

$$I = \frac{V_{ph}}{\omega L} = 0.500 \text{ kA}$$

Rating =V_{ph}× I × 1000 = 66600 kVA **Answer**

Prob.4.2.13 Overhead line-Corona

A 3-phase overhead line consists of stranded copper conductors, spaced 3.66 m in in equilateral formation. The diameter of each conductor is 0.94 cm. For disruptive critical voltage, the surface factor is 0.9, and irregularity factor is 0.95; while for "local" and "general" visual corona voltages, the surface factors are 0.7 and 0.8 respectively. If the ambient temperature is 26.67 deg. C, and the barometric pressure is 74.2 cm Hg, calculate

(a) the voltage at which corona begins
(b) the voltage for "local" corona
(c) the voltage for "general" corona
The dielectric strength for air is 21.1 kV (r.m.s.) per cm at N.T.P

Solution. (a)

The voltage at which corona loss begins is called "disruptive critical voltage", E_o

$$E_o = 21.1\ m\ \delta r \left(1+\frac{0.3}{\sqrt{\delta r}}\right) \ln\left(\frac{d}{r}\right)\ \text{kV (r.m.s.)}$$

$\delta = 3.92\ b/\ (273+T)\ b = 74.2$ cm

$T = 26.67°C$

$\delta = 3.92\ b/\ (273 + T) = 0.9706$

m_d surface factor for disruptive voltage = 0.9

m_i Irregularity factor = 0.95

$m = m_d \times m_i = 0.855$

d, spacing = 366 cm

r, radius = 0.47 cm

$$E_o = 21.1\ m\delta r \left(1+\frac{0.3}{\sqrt{\delta r}}\right) \ln\left(\frac{d}{r}\right)\ \text{kV (r.m.s.)} = 79.128\ \text{kV (r.m.s. to earth)} = 136.89\ \text{kV (L-L)}$$

Answer

(b)

When the voltage is increased beyond E_o, corona begins to appear first in patches, which is called local corona, and the voltage at which this phenomenon takes place is called E_v, Visual Critical Corona. If the voltage is increased still further, increased corona appears along the whole conductor quite uniformly. The voltage at which this 'general' corona takes place is E_v'

m_d surface factor for local corona = 0.7

m_i Irregularity factor = 0.95

$m = m_d \times m_i = 0.665$

$$E_v = 21.1\ m\ \delta\ r \left(1+\frac{0.3}{\sqrt{\delta r}}\right) \ln\left(\frac{d}{r}\right)\ \text{kV (r.m.s.)} = 61.544\ \text{kV r.m.s. to earth}$$

= 106.472 kV (L – L)

Answer

m_d surface factor for "general" corona = 0.8

m_i Irregularity factor = 0.95

$m = m_d \times m_i = m_d{*}m_i = 0.76$

$$E_v' = 21.1\ m\ \delta\ r \left(1+\frac{0.3}{\sqrt{\delta r}}\right) \ln\left(\frac{d}{r}\right) \text{kV (r.m.s.)} = 70.336\ \text{kV rms to earth} = 121.682\ \text{kV (L-L)}$$

Answer

Prob.4.2.14 Overhead lines-Performance using nominal T representation

A balanced load of 50 MW at .8 power factor lagging is supplied at 132 kV, 50 Hz by means of a transmission line. The series impedance of a single conductor is (20+ j 50) ohms and the total phase-neutral admittance is 300 micro Siemens. Using the nominal T approximation, calculate the sending end voltage. If the load is removed and the sending end voltage remains constant find the percentage rise in voltage at the receiving end.

Solution.

$R = 20\ \Omega$

$X_l = 50\ \Omega$

$Y = 0.0003$ S

f frequency = 50 Hz

$\omega = 2\pi f = 376.8$ rad/s
Re **Z** = 20 Ω
Im **Z** = 50 Ω
Im **Y** = 0.000 S
A = 1 + ZY/2
ReA = 1 + (ImZ × ImY)/2 = 0.993
Im A = – ReZ × ImY/2 = 0.003
B = Z × (1 + ZY/4)
Re (1 + ZY/4) = 1 + (ImZ × ImY)/4 = 0.996
Im (1 + ZY/4) = – ReZ × ImY/4 = 0.0015
Re B = 19.85
Im B = 50
P power = 50000000 W
Vr per phase = 76210 Volts
p_f = 0.8 lag
I_r = P/ (3 × V × pf) = 273.37 A
Re I_r = 218.69 A
Im I_r = – 164.02 A
Vs = A × Vr + B × Ir Vr: Reference
ReV_s = ReA × V_r + Re B × ReI_r – Im B × ImI_r = 79472.80 Volts
ImV_s = Im A × V_r + Re B × ImI_r + ReI_r × Im B = 7873.05 Volts
$V_s$mag = 79861.82 Volts/ph
= 138324.66 Volts (L-L) **Answer**
Load removed, **Vs = A*Vr**
Vr_{mag} = Vs_{mag}/A_{mag}
139369.30 Volts (L– L) **Answer**
Regulation = ((Vr no load-Vr load)/Vr load) × 100 = 5.71% **Answer**

Prob.4.2.15 Overhead lines - Potential distribution across a string of insulators-Example 3

A post type insulator consists of three pin insulators fixed one above another and used to support a bus bar of a 115 kv 3-phase system, as shown in Fig.4.2.15:

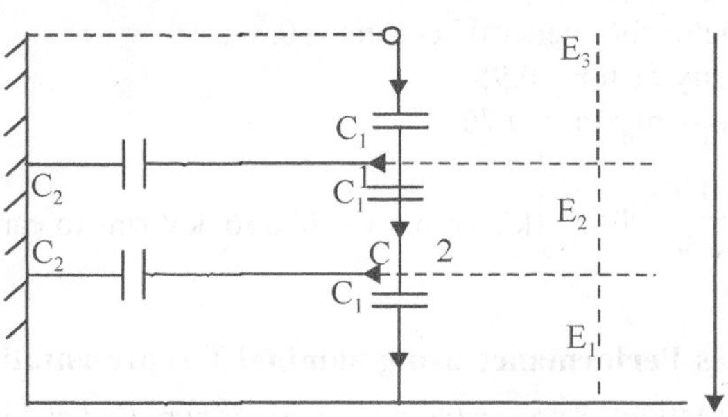

Fig.4.2.15. *Insulator string*

If the voltage across the top pin insulator is twice that of the voltage across the bottom insulator, determine the voltage across the middle insulator.

Solution.

Writing KC_L at Node 1:
$E_2.C_1 = E_3.C_1 + (E_1 + E_2).C_2$
Writing KCL at Node 2:
$E_1.C_1 = E_2.C_1 – E_1.C_2$

Rewriting the two equations:

$E_1.C_2 + E_2.(C_2 - C_1) + E_3.C_1 = 0$ . . . (1)

$E_1.(C_1 + C_2) - E_2.C_1 = 0$ . . . (2)

$E_1 + E_2 + E_3 = E$, say . . . (3)

We have to solve the above equations:

Given that $E_3 = 2.E_1$, the equations are written as:

$3E_1 + E_2 = E$ . . . (i)

$(C_1 + C_2)E_1 - C_1.E_2 = 0$ . . . (ii)

$(C_2 + {}_2.C_1).E_1 + (C_2 - C_1).E_2 = 0$ . . . (iii)

Let $C_1/C_2 = k$, then

$3E_1 + E_2 = E$ . . . (i)

$(k + 1)E_1 - kE_2 = 0$ . . . (ii)

$(1+2.k)E_1 + (1 - k).E_2 = 0$ . . . (iii)

Eliminate E_1 from eq.(i) & eq.(ii) and obtain

$(k + 1).(E - E_2)/3 - kE_2 = 0$ . . . (iv)

Eliminate E_1 from eq.1 & eq.(iii) and obtain

$(1 + 2k)(E - E_2) / 3 + (1 - k) E_2 = 0$ . . . (v)

Eliminate k from eqs. (iv) & (v), and obtain the quadratic:

$3E_2^2 - EE_2 + E.E = 0$ . . . (vi)

Given that,

E (L-L) = 115 kV

E phase value = 66.395 kV

Solving the quadratic Eq.(vi), we get

$E_2 = (3E - \sqrt{5}\,E)/2 = 25.360$ kV **Answer**

Prob.4.2.16 Overhead lines - Receiving end power circle diagram

A 50 Hz 3-phase transmission line is 175 km long. It has line constants A = 0.936 ∠ 0.95° and B = 141.1∠76.25° (ohms). It delivers 40 MW at 220 kV with 90 %pf lagging. Construct a receiving end power circle diagram for this line. Locate the point corresponding to the load and locate the centre of circles for various values of V_s if V_r = 220 kV Draw the circle passing through the load point. From the measured radius of the latter circle, determine V_s.

Solution. Amag = 0.936

Vr L-L = 220.000 kV

B_{mag} = 141.1 Ω

$A_{mag} \times Vr^2/B_{mag}$ = 321.065 MVA

α angle of A = 0.95°

β angle of B = 76.25°

α angle of A = 0.0165 rad

β angle of B = 1.330 rad

Centre (x, y) of circle :

$x = -\ AVr^2.\cos(b - a)/b = -81.679$ MW

$y = -AVr^2.\sin(b - a)/b = -310.50$ MVAr

p.f. = 0.9 lag

p.f. angle = 25.855°

Centre of circles is at – 81.8 MW,–311 MVAr

Construct load line through origin at $\cos^{-1}(0.9) = 25.9\ 0°$.

Draw vertical line at 40 MW. Load point is at intersection of this line and load line in the first quadrant.

Use scale of 1 cm = 25 MVA

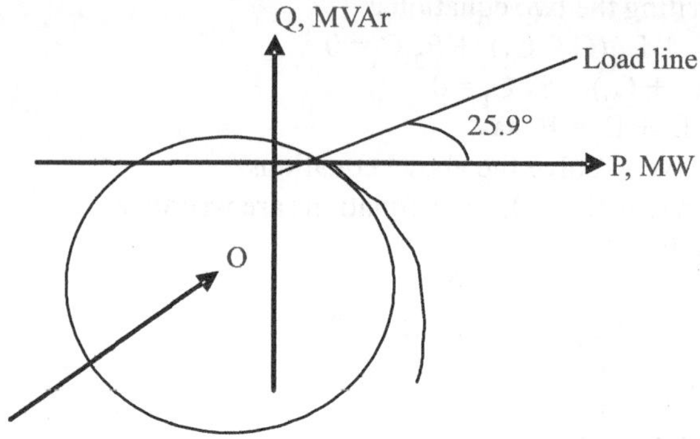

Fig.4.2.16. *Power Circle Diagram*

Radius of circle through the load point is 14.08 cm which is equivalent to 14.08 × 25 = 352 MVA

$$\frac{Vs.Vr}{B} = 352$$

Therefore , Vs = 225.76 kV **Answer**

Prob.4.2.17 Overhead lines -Plot of receiving end voltage against power

A load is supplied through a 275 kV link of total reactance 50 Ω from an infinite busbar at 275 kV. Plot the receiving end voltage against power for a constant load pf of 0.95 lagging.

Solution.

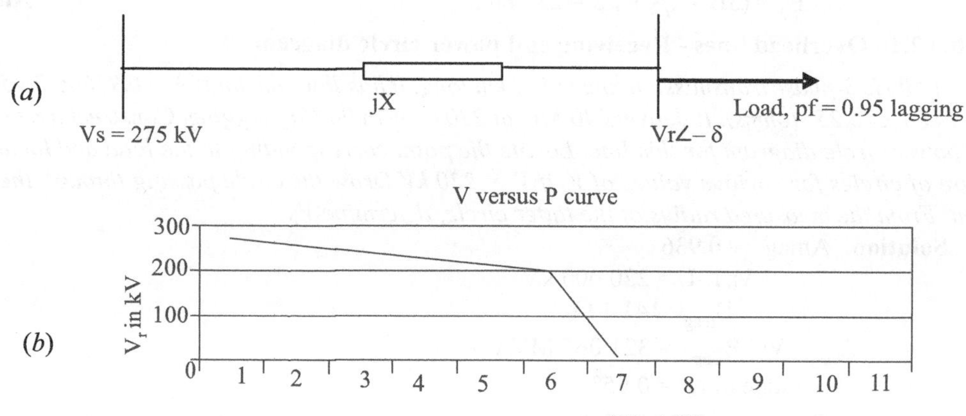

Fig.4.2.17. (*a*) *System diagram* (*b*) *Voltage-Power Curve*

$P = V_s . V_r . \sin(\delta)/X$

$$Q = \frac{V_s . V_r . \cos\delta - V_s^2}{X}$$

Eliminate δ from the above two equations and obtain

$aVr^4 + bVr^2 + c = 0$ where

$a = 1$

$b = 2QX - V_s^2$

$c = P^2X^2 + Q^2X^2$

Solve the above quadratic equation for various values of P in steps of 33.33 MW starting from 0 MW up to 700 MW.

Q = P tan ϕ. Draw the curve shown in Fig.4.2.17 (b).

CHAPTER 5

UNDERGROUND CABLES

5.1 Introduction

A considerable amount of transmission & distribution, especially in urban areas is carried out by means of underground cables. In order to preserve amenities of both town and countryside the electricity supply authorities resort to underground transmission & distribution. Underground transmission is more expensive than the overhead alternative.

5.1.1 Insulating Materials

Dielectric properties of cable insulation :

1. high insulation resistance
2. high dielectric strength
3. good mechanical properties
4. immune to attacks by acids & alkalis
5. non-hygroscopic

Commonly used insulating materials :

(*a*) Oil-impregnated paper
(*b*) Vulcanized India rubber (V.I.R)
(*c*) Polyvinyl chloride (P.V.C)
(*d*) SF_6 gas
(*e*) Cross-linked polythene (XLPC)

Void formation:

Voids (small pockets of air or gas) are formed in the insulation where constituent parts of the cable are expanded and contracted to different extents with heat evolved on load cycles. The stress across the voids is high and breakdown results.

5.1.3 Types of cables

1. Single-core cable
2. Three-core cable
   (*a*) Belted -type construction
   (*b*) H-type construction
3. Oil-filled cable
4. Gas-filled cable- consisting of a conductor supported in a rigid external pipe which is filled with a gas under pressure- usually SF_6 at 3 atmospheric pressure
5. XLPE cables

5.1.4 Laying of cables

(*a*) Direct in the soil
(*b*) In ducts or troughs
(*c*) In circular ducts or pipes
(*d*) In air

5.1.5 Electrostatic stress in a single core cable

The maximum potential gradient occurs at the surface of the conductor. Potential gradient at any point at a distance x from the centre of the conductor is

$$G = \frac{V}{x \log_e \left(\frac{R}{r}\right)}$$

where R is the inner radius of the sheath and r is the radius of the conductor.

Grading of cables

(*a*) Capacitance grading

(*b*) Intersheath grading

5.1.6 Dielectric loss and loss tangent of a cable

The dielectric loss, due to leakage and hysteresis effects in the dielectric, is usually expressed in terms of the loss angle δ:

$\delta = 90 - \phi$

where ϕ is the dielectric power factor angle.

Dielectric loss $= \omega C V^2 \tan \delta$,

where,

C = capacitance to neutral

V = phase voltage

A typical value of $\tan \delta$ lies in the range 0.002 to 0.003. In low voltage cables the dielectric loss is negligible, but is appreciable in EHV cables.

5.1.7 Cables for D.C transmission

Owing to the absence of periodic charging currents with direct voltage, high voltage cables will play an increasingly important role in D.C transmission links. In A.C cable a power factor of 0.003 can be represented by a loss resistance of $3 \times 10^{12}\ \Omega$-cm. The D.C resistivity of the same dielectric would be greater than 10^{14} ohm-cm. Hence the loss in the dielectric on D.C. is only about 3 % of that on A.C. Whereas the electric stress distribution in A.C cables is determined by the dielectric capacitance, in D.C cables it is determined by the electric resistance of the dielectric. The electric resistivity of the conventional dielectrics is very temperature dependent; for oil-impregnated cable, for example, the resistivity at 20°C is 100 times that at 600°C. In D.C cable, thermal considerations not only determine the rating but also influence the electric stress distribution in the dielectric. The electrical resistivity also varies with the electric stress. Instead of electric stress decreasing through the dielectric from the conductor to the sheath, in D.C cables the stress increases and can be larger at the sheath. This is known as stress inversion and can lead to troubles at terminations and joints where the longitudinal stresses are created.

Explain the phenomenon of void formation in cables? Why is the void subjected to excessive potential gradient ?

Void formation does not take place in oil-filled cable. Why ?

5.1.8 Heating of cables

The temperature rise of cable depends on the following factors :

1. The production of heat within the external periphery of the cable.
2. The conveyance of the heat as far as the periphery - that is, up to the boundary of the surrounding medium
3. The conveyance of the heat through this medium, and therefore away from the cable.
4. The current rating of the cables.
5. The nature of the load, i.e. whether continuous or intermittent; not infrequently the rating under short-circuit conditions has to be considered.

Heat production

Within the cable, there are three sources of heat :

1. I^2R loss in conductors
2. Dielectric loss
3. Sheath & armour loss

What is the equivalent circuit for calculating sheath losses ?

Why is cross bonding of sheaths done ?

5.1.9 Current carrying capacity of cables

The limiting factor in current rating is the temperature to which the insulation nearest the conductor can be raised without suffering deterioration.

The allowable values of temperature rise for different types of cables may be obtained from the manufacturer's data books.

The current rating I of a cable neglecting dielectric losses is given by

$$I = \sqrt{\frac{\theta - \theta_a}{nR(S_1 + (1+\lambda)(S_2 + G))}} \text{ Amp.}$$

where,

θ = Core temperature

θ_a = ambient temperature

n = number of conductors

R = Resistance of each conductor

λ = Sheath loss/core loss

S_1 = thermal resistance of the dielectric

S_2 = thermal resistance of the protective covering

G = thermal resistance of the ground

- Write down the expressions for computing the various thermal resistance components.
- What is the effect of temperature on dielectric loss? Modify the formula for current rating considering the effect of dielectric loss
- Discuss the factors affecting the short-circuit rating of an underground cable
- Name the types of cable used for different voltage levels

5.1.10 Tests on Electrical Materials

Type Tests – Tests carried out to prove conformity with the specifications. These are intended to prove the general qualities and design of a given type of manufactured item.

Routine Tests-Tests carried out on each part/item manufactured to check parameters (as per requirements), which are likely to vary during production.

Acceptance Tests- Tests carried out on samples taken at random from offered lot of manufactured item for the purpose of acceptance of lot.

PVC INSULATED CABLES up to and including 1.1 kV [IS: 1554(part-1)-1988]

TYPE TESTS

No.	*Type test*	*Purpose*
(a)	Tests on conductor.	
	1. Annealing test (for copper)	To check softness of wire
	2. Tensile test (for aluminium)	To check strength of Al wire
	3. Wrapping test (for aluminium)	To check hardness of Al wire
	4. Resistance test	To check cross-section of the conductor
(b)	Tests for armouring wires/strips	To check electrical , mechanical and chemical properties of armouring wire/strip

(c) Test for thickness of insulation and sheath	To check capability of insulation to withstand voltage and its mechanical strength
(d) Physical test for insulation & sheath	To check mechanical stress and strain during manufacturing
1. Tensile strength & elongation at break	and bending
2. Ageing in air oven	To check physical & chemical changes in insulation due to heat with age
3. Shrinkage test	To prevent problem in termination
4. Hot deformation	To check resistance against deformation due to heat & mechanical pressure
5. Loss of mass in air oven	To check physical & chemical changes in insulation due to heat and time
6. Heat shock test	To check ability of cable against overheating
7. Thermal stability	To check thermal effect
e) Insulation resistance test	To check uniformities of insulation in dielectric
f) High voltage test(Water immersion test)	To check ability of cable in water during service
g) High voltage test at room temperature	To check ability of cable against high voltage during service
h) Flammability test	To check flame retardant properties

OPTIONAL TYPE TESTS

No. *Optional type test*	*Purpose*
a) Cold bend test	To check effect of low temperature during bending
b) Cold impact test	To check effect of low temperature on outer sheath in terms of hardness & softness
c) Armour resistance test (for other than mining cables)	To check electrical properties of armouring wire/strip

ROUTINE TESTS

No. *Routine Test*	*Purpose*
a. Resistance test	To check cross-section of the conductor
b. High voltage test at room temperature	To check ability of cable against high voltage during service
c. Armour resistance test (for mining cables)	To check conductivity of armouring materials

ACCEPTANCE TESTS

The following type tests are taken as acceptance tests: Type test Nos. a1, a2, a3, a4, c, d1, e, and g

SCALE OF SAMPLING

No. of Drums in a Lot	*No. of Drums to be taken as sample*	*Permissible No. of Defectives*
Up to 50	2	0
51 to 100	5	0
101 to 300	13	0
301 to 500	20	1
501 and above	32	2

CROSS –LINKED POLYETHYLENE INSULATED PVC SHEATHED CABLES [XLPE from 66 to 220 kV] [IS: 7098 (Part-3)-1988]

TYPE TESTS:

No.	Type Test	Purpose
a)	Tests on conductor	To check softness of wire
1.	Annealing Test (for Cu)	
2.	Resistance Test	To check cross-section of the conductor

Test	Purpose
b) Physical tests on insulation	
1. Test for thickness & dimensions of insulation	To check capability of insulation to withstand voltage and its mechanical strength
2. Tensile strength & elongation at break	To check mechanical stress & strain during manufacturing & bending
3. Thermal ageing in oven	To check physical & chemical changes in insulation due to heat with age
4. Hot set test	To check cross-linking of insulating material
5. Shrinkage test	To prevent problem in termination
6. Void & contaminants test	To check voids & contaminants
c) Resistivity test for semi conducting layers	To check resistance of semi-conducting layer
d) Test for concentric metallic screen	To check capacity against short circuit
i) Test for concentric metallic screen	
ii) Test for concentric copper tape	
e) Thickness of metallic sheath	To check capability of insulation to withstand voltage and its mechanical strength
f) Tests for armouring material	
1. Dimensions	To check that dimensions are within limits
2. Tensile strength & elongation at break	To check mechanical stress and strain during manufacturing and bending
3. Wrapping test	To check mechanical strength during bending
4. Resistivity test	To check resistance of armoring material
g) Physical tests for outer sheath	
1 Measurement of thickness	To check mechanical strength
2 PVC Sheath	To know the material used
1. Tensile strength & elongation at break	To check mechanical stress and strain during manufacturing and bending
2. Thermal Ageing in air oven	To check physical & chemical changes in sheath due to heat with age
3. Loss of mass	To check physical & chemical changes in insulation due to heat and time
4. Heat shock test	To check ability of cable against overheating
5. Hot deformation test	To check resistance against deformation due to heat & mechanical pressure
6. Shrinkage test	To prevent problem in termination
7. Thermal Stability	To check thermal effect
3. Polythelene insulated SHEATH	To know the material used
1. Carbon black content	To know the % of carbon
2. Tensile strength & elongation at break before & after ageing	To check mechanical stress and strain during manufacturing and bending
3. Hot deformation	To check resistance against deformation due to heat & mechanical pressure
h) Flammability test (for PVC outer sheathed cable only)	To check flame retardant properties
j) Water tightness test	To check penetration of water in cable
k) 1. Thermal ageing on complete cable sample	To check physical &chemical changes in cable due to heat with age
2. Tensile strength & elongation at break for insulation & outer sheath	To check mechanical stress and strain during manufacturing and bending
3. Resistivity test for semi- conducting layers	To know resistance of semi-conducting layer
m) Bending test followed by P. D. test	To check bending radius during bending while installation & handling
n) Dielectric power factor measurement at ambient temperature	To check rupturing capacity & voids
p) Dielectric power factor measurement at elevated temperature	To check impurities & voids
q) Load cycle test followed by P.D measurement	To check capacity of cable under loading conditions
r) Impulse withstand test followed by HV test	To check ability of insulating material to withstand lightning voltage

NOTES: Tests from (n) to (r) shall be performed successively on the same test sample of complete cable, not less than 10 m length between test accessories.Tests at (p) and (q) may be carried out on different samples.

OPTIONAL TYPE TEST

No. Type Test	Purpose
1. Cold impact test for outer sheath	To check effect of low temperature on outers heath in terms of hardness & softness

ROUTINE TESTS

No. Routine Test	Purpose
a Conductor resistance test	To check cross-section of the conductor
b P. D. test	To check small voids and cavities in insulation
c HV test	To check ability of cable in service

ACCEPTANCE TESTS

No.	Acceptance Test	Purpose
1	Measurement of capacitance	To check impurities & voids

The following Type Tests will be used as Acceptance Tests:

$a_1, a_2, b_1, b_4, b_6, e, g_1$

The following Routine Tests will be used as Acceptance Tests:

b, c

Partial discharge test shall be carried out on full drum length

SCALE OF SAMPLING

No. of Drums in a Lot	No. of Drums to be taken as Sample	Permissible No. of Sample
Up to 25	3	0
26 to 50	5	0
51 to 100	8	0
101 to 300	13	1
301 and above	20	1

DRUMS FOR ELECTRIC CABLES [IS: 10418-1982]

The tests under TYPE, ROUTINE and ACCEPTANCE categories are not specified in the Indian Standards. However, the following checks shall be made on DRUMS & their components.

CHECKS FOR CONSTRUCTION OF DRUM

S. No.	Description	Purpose
1.	Mechanical strength	(a) Transverse loading test (b) Impact test (c) Barrel batten test
2.	Flange & outside surface	Free from protruding materials or Projections or unevenness capable of damaging the cable/hands
3.	Flanges (Main Discs) construction	(a) For dia. up to 1600mm- 2 ply OR 3 ply construction. (b) For dia. Above 1600 mm- 2 full ply OR 3 full ply plus 1 segmental layer construction (Segments shall not be less than six)
	Width of middle plank (Minimum)	For flange dial up to 700mm-100mm Dia 701 mm-1600mm-150 mm Dia above 1600 mm- 200 mm
4.	Barrel end-supports	Shall be complete circular discs or of various segments. Securely fixed to inside of flanges by nailing
5.	Barrel middle-supports	Shall be complete circular construction of single/two ply layers (at 90°) OR of various segments (Only for drums having transverse above 1000 mm).
6.	Stretchers (Core carrier planks)	To be provided for drum sizes of 1206 mm and above

7. Tolerances in mm	mm
Drum flange dial, up to & including 1600 mm	+/- 20
Above 1600 mm	+/-30
Flange thickness up to & including 1600 mm	+/- 06
Above 1600 mm	+/-09
Barrel dia. up to & including 1600 mm	+/- 20
Above 1600 mm	+/-30
Overall & transverse widths	+/- 25
Barrel battens thickness	+/-3
Stretchers thickness	+/-3
Centre hole dia. with bush	0 to+2
Centre hole dia. without bush	0 to +5

Other standards are :

CROSS –LINKED POLYETHYLENE INSULATED PVC SHEATHED CABLES [XLPE up to 3.3 kV] [IS: 7098 (Part-1)-1988]

CROSS –LINKED POLYETHYLENE INSULATED PVC SHEATHED CABLES [XLPE from 3.3to 33 kV] [IS: 7098 (Part-2)-1988]

OBJECTIVE TYPE QUESTIONS

1. Void formation occurs in

(a) XLPE cables (b) Oil-filled cables

(c) Oil-impregnated paper cables

Ans. (c)

2. Insulation resistance of a cable 20 km long is 1 MΩ. Two cable lengths, 20 km and 10 km, are connected in parallel. The insulation resistance of the parallel combination is

(a) 1 MΩ (b) 0.5 MΩ

(c) 0.666 MΩ

Ans. (c)

3. If the voltage applied to the core and sheath of a cable is halved , the reactive power generated by the cable will be

(a) Halved (b) 1/4 th of the original value

(c) doubled

Ans. (b)

4. The dielectric field intensity at a point within the dielectric of a cable

(a) is constant

(b) increases with increase of distance of the point from the centre of the cable

(a) decreases with increase of distance of the point from the centre of the cable

Ans. (c)

5. Three insulating materials with identical maximum working stress and permittivities of 2.5, 3 and 4 are used in a single-core cable. The location of the materials with respect to the cable core will be

(a) 2.5,3,4 (b) 3,2.5,4

(c) 4,3,2.5 (d) 4,2.5,3

Ans. (c)

6. Three insulating materials with breakdown strengths of 2.5,3 and 3.5 are used in a single-core cable. If the factor of safety for the materials is 5, the location of the materials with respect to the core of the cables will be

(a) 2.5,3,3.5 (b) 3,2.5,3.5

(c) 3.5,3,2.5 (d) 3.5,2.5,3

Ans. (c)

7. If δ is the loss angle of a cable , its power factor is
   (a) sinδ
   (b) cosδ
   (c) p .f. is independent of δ
   (d) p. f depends on δ but not as in ' a' or 'b'
   Ans. (a)
8. **Match List A with List B.**

List A	**List B**
Voltage range	Critical design factor for cable
I. up to 33 kV	p. Thermal instability
II. 33-132 kV	q. Ionization
III. above 132 kV	r. Impulse strength

   The correct matching is
   (a) Ip IIq IIIr
   (b) Iq IIp IIIr
   (c) Iq IIr IIIp
   Ans. (c)
9. If C_1 is the capacitance between any two cores of a 3-core cable, and C_2 is the capacitance between any core and the sheath, then the measured value of the capacitance between any two cores with the third core isolated is equal to
   (a) $C_1C_2/(2C_1+C_2)$
   (b) $0.5(3C_1+C_2)$
   (c) $3C_2$
   Ans. (b)
10. Sheaths are provided in cables to
   (a) provide proper insulation
   (b) provide mechanical strength
   (c) prevent ingress of moisture

PROBLEMS WITH SOLUTIONS

Prob.5.2.1 Underground cable-3-core-Tests to determine capacitances

In a test of a 3-conductor underground cable, shown in Fig.5.2.1,

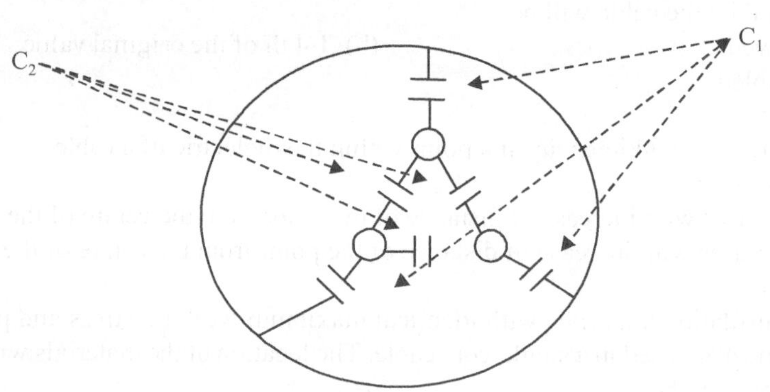

Fig. 5.2.1. *Cable capacitances*

the 3 conductors are bunched together, and the capacitance between the bunched conductors and the sheath is found to be C_a. The capacitance between two of the conductors bunched together with the sheath, and the third conductor is found to be C_b. Determine C_1 and C_2 for the cable, in terms of C_a and C_b.

Solution. $C_a = 3C_1$

$C_b = C_1 + 2*C_2$

Therefore,

$$C_1 = \frac{1}{3}Ca$$ **Answer**

$$C_2 = \frac{1}{2}\left(C_b - \frac{1}{3}Ca\right)$$ **Answer**

Prob.5.2.2. Underground cable-3-core-Charging kVA

In a 3-core cable, the capacitance between the cores short-circuited together and the sheath is 0.54 microfarad/km and that between two cores connected together and the third core is 0.52 microfarad /km. Determine the kVA required to keep 10 km of the cable charged when the supply is 33kV, 3 phase, 50 Hz.

Solution.

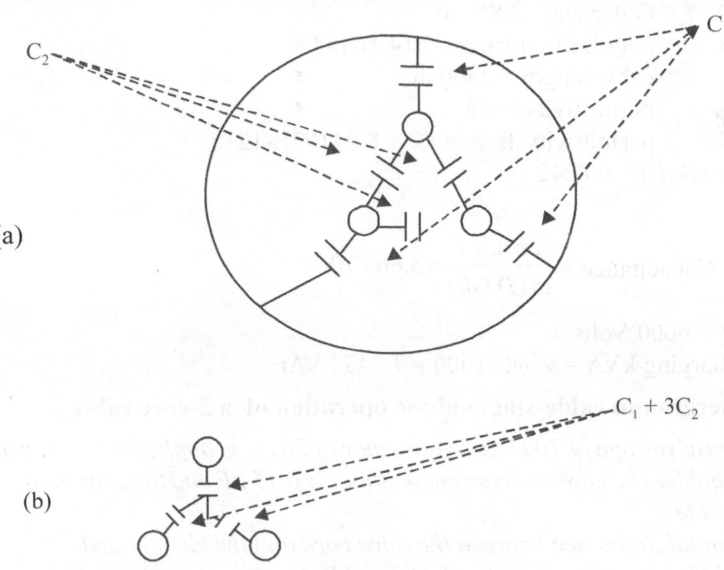

Fig. 5.2.2. (a) Cable capacitances (b) Equivalent circuit: star

C_a = Capacitance between the three cores bunched together and the sheath = 0.54

C_b =capacitance between two cores connected together and the third core = 0.52

From the solution of Prob.5.2.1, we have

$$C_1 = \frac{1}{3}Ca = 0.18$$

$$C_2 = \frac{1}{2}(C_b - \frac{1}{3}Ca) = 0.17$$

Capacitance /phase, $C_0 = C_1 + 3C_2$

$Co = 0.69\ \mu F/km$

$Co = 0.00000069\ F/km$

$V(L\text{-}L) = 33\ kV$

$f = 50\ Hz$

$\omega = 2\pi f = 314.16\ rad/s$

l cable length = 10 km

kVA, 3-phase $= V^2(L-L)\times\omega.Co.l.1000$

= 2360.629 kVA **Answer**

Prob.5.2.3. Underground cable-single core- Charging kVA

A single-phase concentric cable 2.4 km long is connected to 50 Hz, 6.6 kV bus bars. The inner conductor has a diameter of 1.25 cm and the radial thickness of insulation is 0.8 cm. The relative permittivity of the dielectric is 3.5. Determine the charging kVA.

Solution.

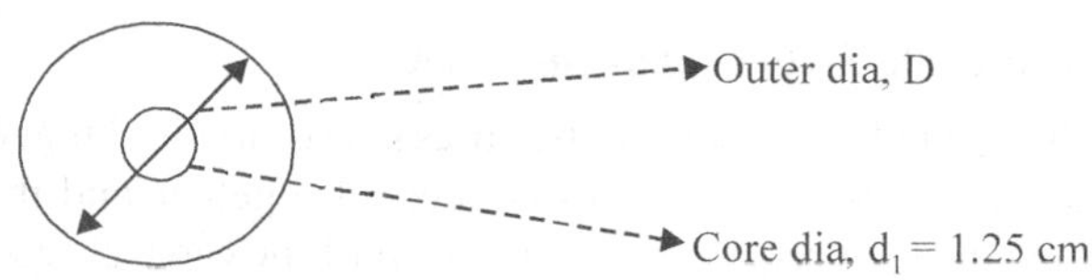

Fig.5.2.3. A single-phase cable

d_1, core dia = 1.25 cm
t, radial thickness = 0.8 cm
D, Outer dia = 2.85 cm
ω, angular frequency = 314.16 rad/s
l, cable length = 2400 m
ε_r, permittivity = 3.5
E_o, permittivity-free space = 8.84192E-12
$\ln(D/d_1) = 0.8242$

$$\text{C Capacitance} = \frac{2\pi\,\varepsilon_0\varepsilon_r l}{\ln(D/d_1)} = 5.66\times10^{-7}$$

V = 6600 Volts
charging kVA = $V^2\omega C/1000$ = 7.745 kVAr **Answer**

Prob.5.2.4 Underground cable-single-phase operation of a 3-core cable

A single-phase voltage of 10 kV at a frequency of 50 Hz is applied between two of the cores of a 3-phase belted cable. The core -to -core capacitance is 0.15 μF and the core -to sheath capacitance is 0.01 μF. Calculate

(a) the potential difference between the third core and the sheath, and
(b) the total charging current taken by the cable.

Solution.

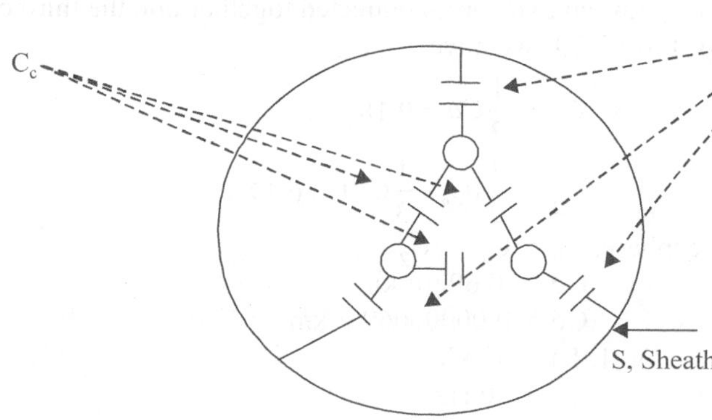

(a)

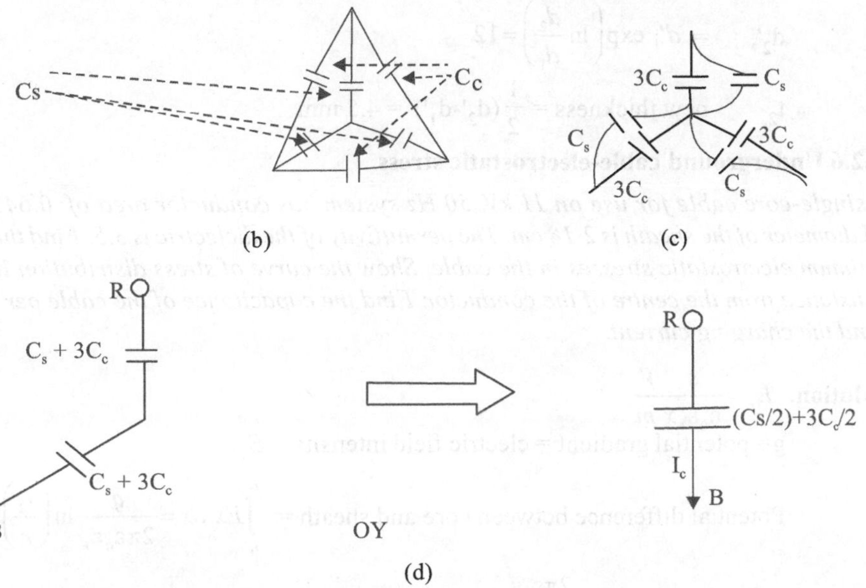

Fig.5.2.4. *(a) Cable (b), (c), (d) Equivalent circuits*

Voltage between Y (third core) and sheath = zero **Answer**

Single-phase voltage is applied between R and B

$V = 10000$ Volts

$\omega = 314.16$

$C_s = 0.0000001$ F

$C_c = 0.00000015$ F

I_c, Current $= V\omega\left(\frac{C_s}{2}+3\frac{C}{2}\right) = 0.863$ A **Answer**

Prob. 5.2.5 Underground cable-single core-Insulation resistance

A cable composed of a conductor of 2 mm diameter covered with 1 mm of insulating material is found to have an insulation resistance of 480 MΩ/km per km. What thickness of a similar material would be required for a 3 mm diameter conductor in order to have an insulation resistance of 960 MΩ/km ?

Solution. d_1, conductor dia = 2 mm

t_1, insulation thickness = 1 mm

d_2, cable dia $= d_1+2t_1 = 4$ mm

R, Insulation resistance $= \frac{\rho}{2\pi}\ln\left(\frac{d_2}{d_1}\right) = 480$ MΩ/km

R_2, New insulation resistance = 960 MΩ/km

d_2', = New cable dia

d_1', New conductor dia. = 3 mm

$$R_2, \quad = R\,\frac{\ln(d_2'/d_1')}{\ln(d_2/d_1)}$$

$$\ln\left(\frac{d_2'}{d_1'}\right) = \frac{R_2}{R}\ln\left(\frac{d_2}{d_1}\right)$$

$$= 1.386$$

d_2', $= d'_1 \exp\left(\ln\frac{d_2}{d_1}\right) = 12$

t_2, new thickness $= \frac{1}{2}(d_2' - d_1') = 4.5$ mm **Answer**

Prob.5.2.6 Underground cable-electrostatic stress

A single-core cable for use on 11 kV, 50 Hz system has conductor area of 0.645 cm². The internal diameter of the sheath is 2.18 cm. The permittivity of the dielectric is 3.5. Find the maximum and minimum electrostatic stresses in the cable. Show the curve of stress distribution in the cable versus distance from the centre of the conductor. Find the capacitance of the cable per km length. Find the charging current.

Solution. $E_x = \frac{q}{\varepsilon_0 \varepsilon_r x} \frac{V}{m}$

g= potential gradient = electric field intensity = E_x

Potential difference between core and sheath = $\int_{r_1}^{r_2} Ex.dx = \frac{q}{2\pi\varepsilon_0\varepsilon_r}\ln\left(\frac{r_2}{r_1}\right)$

$$C = q/V = \frac{2\pi\varepsilon_0\varepsilon_r}{\ln(r_2 r_1)}$$

$$Q = \frac{2\pi\varepsilon_0\varepsilon_r V}{\ln\left(\frac{r_2}{r_1}\right)}$$

$$g = \frac{V}{x\ln\left(\frac{r_2}{r_1}\right)}$$

$$g_{max} = \frac{V}{r_1 \ln\left(\frac{r_2}{r_1}\right)}$$

$$g_{min} = \frac{V}{r_1 \ln\left(\frac{r_2}{r_1}\right)}$$

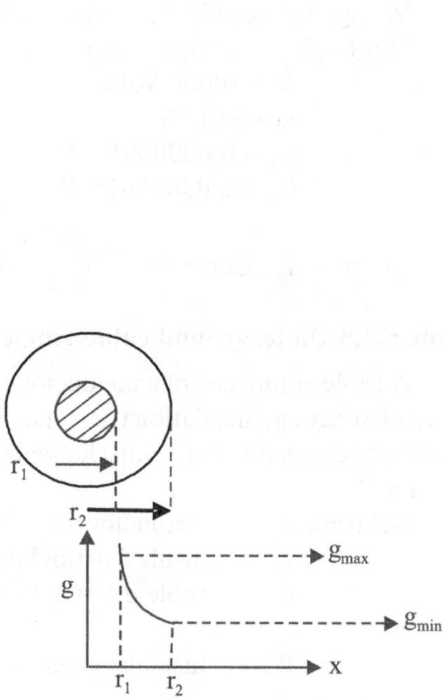

Fig.5.2.6. *Cable with graph of Stress distribution*

V = 11000 Volts

A, conductor area = 0.645 cm²

r_1, conductor radius = 0.453 cm

r_2 = 1.09 cm

g_{max} = 27656.328 V/cm,

g_{min} = 11496.68 V/cm

ε_r = 3.5

ε_o = 8.841E-12

$C = q/V = \frac{2\pi\varepsilon_0\varepsilon_r}{\ln\left(\frac{r_2}{r_1}\right)}$ 2.215×10^{-7} F/km **Answer**

ω = 314

I = ωCV= 0.765 A **Answer**

Prob.5.2.7 Underground Cable-Condition for maximum electric stress to be minimum

Show that for a given voltage, the maximum stress E_{max} in the dielectric of a single - core cable the sheath diameter D is a minimum when D = ed where d is conductor diameter and e is the base of natural logarithms. Determine D and d for V= 10 kV and Emax = 23 kV/cm.

Solution. $g_{max} = \dfrac{V}{r_1 \ln\left(\dfrac{r_2}{r_1}\right)}$

g_{max} is minimum when $r_1 \ln\left(\dfrac{r_2}{r_1}\right)$ is maximum *i.e* when $r_2 = e\, r_1$ or D = e.d

min. $g_{max} = g_{max} = E_{max} = 23$ kV/cm

V = 10 kV

min $g_{max} = V/r_1$

Therefore,

$r_1 = V/\text{min } g_{max} = 0.434$

$d = 2r_1 = 0.869$ cm **Answer**

D = e.d 2.718 × 0.869 = 2.363 cm **Answer**

Prob.5.2.8 Underground cable-Temperature rise of conductor & sheath

Three 132 kV single-core cables rated at 600 A are laid at a depth of 100 cm in horizontal formation 5 cm apart. The thermal resistivity of the soil is 100°C cm/W).The cable details are:

Copper conductor	*3 cm² cross-section*
Insulation	*1 cm in radial thickness*
	Outer radius 2 cm
	g = 500°C cm/W
Lead sheath	*Thickness 0.2 cm*
	g = 3°C cm/W
Covering	*Thickness 0.6 cm*
	g = 500°C cm/W

Rated load copper loss at 85°C is 0.3 W/cm.

Dielectric loss per cm is 0.02 W/cm.

Cables are cross-bonded. Sheath loss is negligible.

Calculate the temperature rise of the conductor and sheath if the ambient temperature is 15°C

Solution.

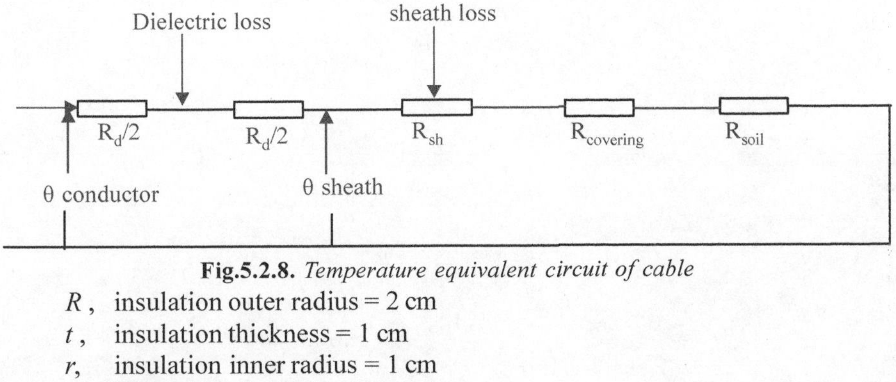

Fig.5.2.8. *Temperature equivalent circuit of cable*

R, insulation outer radius = 2 cm

t, insulation thickness = 1 cm

r, insulation inner radius = 1 cm

R, sheath inner radius = 2 cm

t, sheath thickness = 0.2 cm

R_1, sheath outer radius = 2.2 cm

t_1, covering thickness = 0.6 cm

R_2, cable radius = 2.8 cm
g_1, g of insulation = 500°C-cm/W
g_2, g of sheath = 3°C-cm/W
g_3, g of covering = 500°C-cm/W
g_4, g of soil = 100°C-cm/W

Thermal resistances of dielectric, sheath and covering are R_d, R_{sheath}, and $R_{covering}$

$$R_d = \frac{g_1}{2\pi} ln\left(\frac{R}{r}\right) = 55.158°C/W$$

$$R_{sheath} = \frac{g_2}{2\pi} ln\left(\frac{R_1}{R_1}\right) = 0.0455°C/W$$

$$R_{covering} = \frac{g_3}{2\pi} ln\left(\frac{R_2}{R_1}\right) = 19.191°C/W$$

h cable depth = 100 cm
p cable spacing = 5 cm

Thermal resistance of soil presented to the centre cable = R soil

$$R_{soil} = \frac{g_4}{2\pi} \ln\left(\frac{2h}{R_2}\right) + \frac{g_A}{2\pi} ln\left(\frac{4h^2 + p^2}{p^2}\right) = 185.403°C \ W/cm$$

copper loss = 0.3 W/cm

Conductor temperature rise = Copper loss × (R_d +R_{sh} +$R_{covering}$ + R_{soil}) = 77.939°C

Dielectric loss = 0.02 W/cm

Temp. rise due to dielectric loss = Dielectric loss × (R_d/2) +R_{sh} +$R_{covering}$ + R_{soil}) = 4.644°C

Ambient temperature = 15°C

Conductor temperature = rise due to cu loss + rise due to dielectric loss + ambient temperature
= 97.583°C

CHAPTER 6

TRANSFORMERS

6.1 INTRODUCTION

The *transformer program* in the website (*http://www.msoe.edu/~saadat/matlabgui.htm*)

Menu-Driven Graphical User Interface (Gui) Program by Prof. H. Saadat obtain the transformer equivalent circuit from open-circuit and short-circuit tests. It also finds the transformer performance characteristics using the transformer parameters. In the *induction motor* program the parameters of the equivalent circuit can be entered on the circuit diagram. The program draws the Thevenin's equivalent circuit. Performance characteristics are then obtained for the specified slip and the torque slip curve is plotted on the screen. The analysis can be repeated for different slips. A GUI program deals with the design of a *first-order controller* in the forward path of a closed-loop control system. This GUI program designs the following controllers: Proportional, phase-lag, phase-lead, PD, PI, and PID controllers. To run the controller design program at the MATLAB prompt type >> rl designgui.

6.1.1 Basic studies

Voltage and current ratios, phasor diagrams, losses and efficiency, connections (including auto- and three –winding transformers) and general construction of 1-phase and 3-phase units, leading to the equivalent circuit and voltage-drop calculations.

6.1.2 Specialised problems

Reactance: Leakage-flux paths should be investigated for various winding arrangements, and means of varying the reactance value at the design stage discussed.

Mechanical forces: Associated with the reactance are the axial and radial forces set up with symmetrically-and asymmetrically arranged windings (including the effect of tappings). Cause and nature of these forces should be understood.

Thermal Conditions: Like other electrical plant, a transformer can be given a continuous rating based on its steady-state temperature rise. Transformers, however, usually operate on variable load, as they have a long heating time constant τ, they can be safely overloaded for up to 1 or 2h without exceeding a safe temperature. The relevant calculations are largely based on the temperature /time relation $\theta = \theta_f - (\theta - \theta_i)e^{-t/\delta}$ where θ_f is the final steady, and θ_i is the initial, temperature rise.

Students should be familiar with various methods of natural and forced cooling.

Harmonics: The setting up of voltage and or current harmonics in a 1-phase transformer as a result of the inevitable saturation in the magnetic circuit should be well understood. Similar conditions in a three-phase transformer are modified by the winding connections and the construction of the core(*e.g.*, 3 or 5 limbs) and may lead to an oscillating neutral or to circulating triplen harmonic currents in delta-connected windings. Harmonics and asymmetry in the magnetising inrush current should also be considered.

Noise: Harmonics may result in noise, but magnetisation is a more likely cause: the noise may be amplified by mechanical resonances in core and tank; provided these are minimised as far as possible, and the flux density is not very high, there is little that can be done to eliminate the noise at its source and some form of screening or enclosure must be adopted.

Tap changing: The principals of reactor and resistor schemes for on-load tap changing should be known.

Impulse testing: The vulnerability of a transformer to impulse voltages (lightning etc.) on an H.V network makes this very important. Students should be familiar with the principle and analysis of the Marx circuit for an impulse –testing generator; they should also understand the behaviour of a transformer winding when subjected to an impulse voltage and be aware of the methods for achieving a fairly uniform voltage distribution.

6.1.3 Power Transformers-Indian Standards

Power Transformers [ISD: 2026 (Part –1)-1977]

TYPE TESTS

S. No.	*Type test*	*Purpose*
1	Measurement of winding resistance	To know resistance of winding utilized
2	Measurement of voltage ratio and check of voltage vector relationship	To know correct voltage ratio at each tap and to check vector symbol (To ascertain phase displacement)
3	Measurement of impedance voltage/ short-circuit impedance (principle tapping) and load loss	To know copper (load) loss and impedance voltage
4	Measurement of no-load loss and current	To know core loss & excitation current
5	Measurement of insulation resistance	To check insulation strength
6	Dielectric tests (Lightning impulse on line terminals	To verify impulse voltage withstand capacity of insulation (Line terminal to earth & other winding)
7	Temperature rise test	To check temperature rise of transformer under loading conditions
8	Tests on on-load tap changer (where appropriate)	To ensure proper operation of OLTC

SPECIAL TESTS

S. No.	**Special test**	**Purpose**
1.	Dielectric test (Modified impulse test on Neutral	To verify impulse voltage withstand capacity of insulation (Neutral terminal to earth & other winding)
2.	Measurement of zero sequence impedance of 3-phase transformers transformers	To check zero sequence impedance of 3-phase transformers
3.	Short circuit test condition	To ensure mechanical strength under short circuit
4.	Measurement of acoustic noise level	To measure sound level produced by transformer during rated excitation under no-load condition
5.	Measurement of harmonics of the no-load current	To measure harmonics of the no-load current
6.	Measurement of power taken by fans & oil pumps	To know power consumption of fans & oil pumps

Routine Tests

S. No.	Routine test	Purpose
Type Tests 1,2,3,4,5,6, and 9 will also be the Routine Tests. In addition, the following Dielectric tests will be performed as a Routine test		
Dielectric Tests· • Separate AC source • Induced over voltage		• To verify power frequency withstand capacity of winding (winding to other winding and to earth)· • To verify power frequency withstand capacity of winding (phases to earth and other winding)

ACCEPTANCE TESTS

The routine tests on each transformer are to be considered as acceptance tests.
SAMPLE CRITERIA: Each power transformer is to be tested.

OBJECTIVE TYPE QUESTIONS

1. For a 3-phase transformer, which of the following statements is true?
(a) Turns ratio = line voltage ratio
(b) Turns ratio = phase-voltage ratio (c) Turns ratio = 3 × phase voltage ratio.
Ans. (a)

2. The use of higher flux density in transformer design
(a) reduces the transformer weight per kVA
(b) it increases the weight/kVA (c) increases the weight/kW
Ans. (a)

3. The conservator is used in a transformer because
(a) it supplies oil to the transformer whenever needed
(b) protects the transformer from damage when oil expands due to rise in temperature
(c) it provides fresh air and cools down the oil.
Ans. (b)

4. Which of the following has the highest efficiency ?
(a) d.c. shunt motor (b) transformer
(c) induction motor (d) synchronous motor
Ans. (b)

PROBLEMS WITH SOLUTIONS

Prob.6.2.1 Three-winding transformer

A constant-voltage source (infinite bus) supplies a purely resistive 5 MW,2.3 kV load, and 7.5 MVA, 13.2 kV synchronous motor having a sub-transient reactance X ″ = 20%. The source is connected to the primary of a 3-winding transformer. The motor is connected to the secondary and the resistive load to the tertiary. Draw the impedance diagram of the system, and mark pu impedance for a base of 66 kV, 15 MVA in the primary.

Transformer data :

Primary:	Star	66 kV,15 MVA		
Secondary:	Star	13.2 kV,10MVA		
Tertiary:	Delta	2.3 kV,5 MVA		
X_{ps} =	7%	15 MVA	66 kV	Base
X_{pt} =	9%	15 MVA	66 kV	Base
X_{st} =	8%	10 MVA	13.2 kV	Base

Solution. **Primary :**

Base MVA = 15
Base kV = 66

Secondary:

Base MVA = 15
Base kV = 13.2

Tertiary:

Base MVA = 15
Base kV = 2.3

X_{ps} pu = 0.07
X_{pt} pu = 0.09
X_{st} pu = 0.08
X'_{st} on common base = Xst × Base MVA/Given MVA = 0.12 pu
$Xp = 0.5(X_{ps} + X_{pt} - X_{st}') = 0.02$ pu

$X_s = 0.5\ (X_{ps} + X_{st}' - X_{pt}) = 0.05$ pu
$X_t = 0.5\ (X_{pt} + X_{st}' - X_{ps}) = 0.07$ pu
Load R = 1 pu, on a base of 5 MVA, 2.3 kV in tertiary
base MVA for load R = 5
Rpu on common base MVA = 3 pu
MOTOR:

X″	=	0.2 pu
Motor MVA	=	7.5
X″ on common MVA	=	0.4 pu

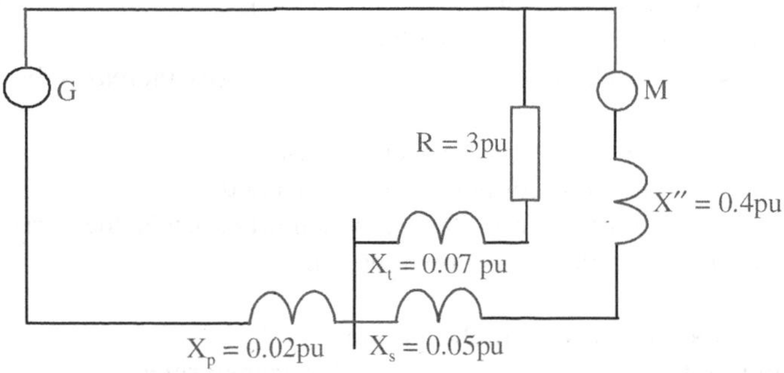

Fig.6.2.1. *Equivalent Circuit-Impedance diagram*

Prob.6.2.2 Transformers Parallel Operation

A generator can develop 600 kW of power. Its voltage of 415 V(L-L) is stepped up to 22 kV by two 3-phase transformers with equal ratios of transformation, and each star-connected on both primary & secondary sides. They operate in parallel to supply a load. Transformer A has a primary current of 500 A and its p.f. on the primary side is 0.7 lagging. Calculate the p.f. and current on the primary side of transformer B. Calculate also the pf of the load and the total load current.

$R_a = R_b = 6\ \Omega$, $X_a = X_b = 16.75\ \Omega$.

These parameters are all referred to the primary and are those for each phase.

If the primary line voltage is 415 V, calculate the resistance and reactance per phase of single star-connected load that would be equivalent to the actual load.

Solution.

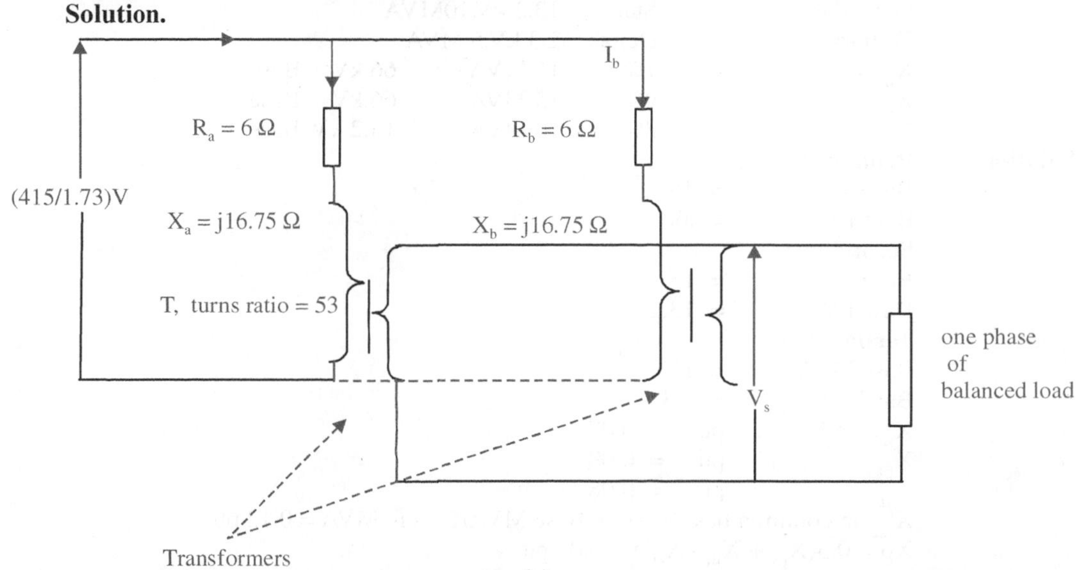

(a)

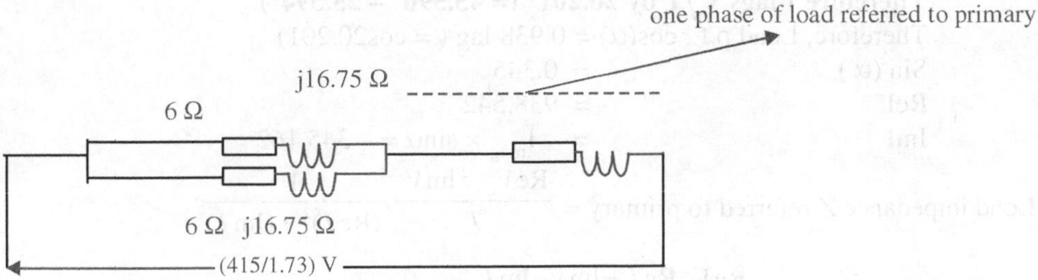

(*b*)

Fig. 6.2.2. (*a*) *Equivalent Circuit of parallel transformers* (*b*) *load.*

V_p / ph = 239.884 Volts **(Reference Voltage)**

V_s / ph = 12716.763 Volts

T, Turns ratio = 53

$\cos\phi$ = 0.7

$\sin\phi$ = = $\sin(\cos^{-1}\phi)$=0.714

I_a mag = 500 A

Re I_a = I_a mag $\cos\phi$ = 350 A

Im I_a = I_a mag $\sin\phi$ = 357.071 A

Equating voltage drops;

$I_a(R_a + jX_a) = I_b(R_b + jX_b)$

Therefore,

Re I_b = Re I_a = 350 A

Im I_b = Im I_a = 357.071 A

Since $X_a/R_a = X_b/R_b$, the primary p.f. of B is also 0.7.

Therefore I_a and I_b are in phase.

$I = I_a + I_b$

Re I = 700 A

Im I = 714.142

I_{mag} = 1000

This is the load current referred to primary.

Actual current in load = 18.863 A

The load p.f. is the cosine of the angle between the load current and V_s or between I_{mag} and V_p (= V_s/T.)

V_s referred to the primary is:

$\mathbf{V_s}/T = V_p - \mathbf{I_a}(R_a + jX_a)$

R_a = 6 Ω

R_b = 6 Ω

X_a = 16.75 Ω

X_b = 16.75 Ω

Re($\mathbf{V_s}$/T) = –7841.061 Volts

Im(Vs/T) = – ReI_a X_a + Im$I_a R_a$ – 3720.071 Volts

$\mathbf{V_s}$/T lags V_p by

$$\tan^{-1}\frac{\mathrm{Im}V_s/T}{\mathrm{Re}V_s/T} = 25.394°.$$

I lags Vp by

$\cos\phi$ = 0.7

ϕ = 45.596°.

Therefore I lags V_s/T by 20.201° (= 45.596° – 25.394°)

Therefore, Load p.f., cos(α) = 0.938 lag (= cos20.201)

Sin (α) = 0.345

ReI = 938.542

ImI = $-I_{mag} \times \sin\alpha = -345.162$

$$\text{Load impedance Z referred to primary} = \frac{\text{Re}V_s \, . \, j\,\text{Im}V_s}{T} \cdot \frac{1}{(\text{Re}\,I)^2 + (\text{Im}\,I)}$$

$$\text{Re (Load impedance)} = \frac{\text{Re}V_s . \text{Re}\,I + \text{Im}V_s . \text{Im}\,I}{T \ \ (\text{Re}\,I)^2 + (\text{Im}\,I)^2} - 6.075$$

$$\text{Im (Load impedance)} = \frac{\text{Re}V_s . \text{Im}\,I . \text{Re}\,I}{T \ \ (\text{Re}\,I)^2 + (\text{Im}\,I)^2} = -6.197$$

Each phase of the equivalent star-connected load at the secondary terminals is therefore $(T^2)(\text{ReZ} + j\text{ImZ})$.

T^2.Re Z = – 17072.840

T^2.ImZ = – 17417.780

or, Z = 17072 + j17417 Ω

Prob.6.2.3 Transformer parameters-Star-star & Delta-star connections

(a) A 3-phase transformer has a nameplate rating of 10 MVA, 132 -11 kV with a leakage reactance of 8 %. The transformer connection is star-star. Select a base of 100 MVA and 132 kV on the HV side and determine:

(i) Transformer reactance in pu

(ii) HV side base impedance

(iii) LV side base impedance

(iv) Transformer reactance in ohms referred to HV side

(v) Transformer reactance in ohms referred to LV side

(b) The above transformer connection are now changed to delta on the low voltage side and star on the HV side. Determine:

(i) Turns ratio of windings

(ii) Transformer reactance in Ω referred to LV side

(iii) Transformer reactance in pu referred to LV side

Solution.

MVA	rating	= 10
kV_L	low side	= 11
kV_H	High side	= 132
MVA	base	= 100
base kV	High side	= 132

(i) Xpu = 0.08 pu **Answer**

(ii) Z_{base}(HV) = Base kV × Base kV/Base MVA = 174.24 Ω **Answer**

(iii) Z_{base} (LV) = 1.21 Ω **Answer**

(iv) X Ω referred to High side = Xpu× Z_{base} (HV)

X(HV) = 13.939 Ω **Answer**

(v) X Ω referred to low side = Xpu × Z_{base}(HV)

X(LV) = 0.0968 Ω **Answer**

(b) (I) Turns ratio,T = $(kV_H/1.73)/kV_L$ = 6.936 **Answer**

(ii) X Ω referred to low side = X Ω (referred to High side)/(T^2) = 0.289 Ω **Answer**

(iii) Reactance of equivalent star connection = X ohms referred to low side/3 = 0.096 Ω

Xpu $= X/Z_{base}(HV) = 0.079$ pu **Answer**

If L-L voltages are used,

$$X(LV) = \frac{X(HV)}{(132/11)^2} = 0.0968\ \Omega$$

$$X(LV)\ pu = \frac{X(LV)}{Z_{back}(LV)} = 0.08\ pu$$

Prob.6.2.4 Transformers-Parallel operation-Example 2

Two 6600/440 – V, star-connected 3 – phase transformers , A of 250 kVA and B of 500 kVA, have the following impedances /phase :

$R_a = 0.008\ \Omega$, $X_a = 0.035\ \Omega$, $R_b = .003\ \Omega$, $X_b = 0.019\ \Omega$.

How will they share a load of 600 kVA at a power factor of 0.8 lagging if the load voltage is 440 V ? The kVA and p.f. of each transformer are required.

Solution :

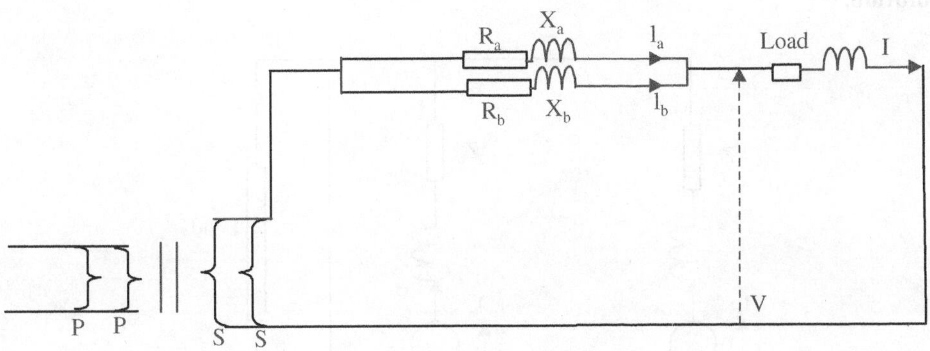

Fig.6.2.4. *Equivalent circuit of parallel transformers*

$\mathbf{I_a = IZ_b/(Z_a + Z_b)}$

$\mathbf{I_b = IZ_a/(Z_a + Z_b)}$

$P_a + jQ_a = \mathbf{V}.\ I_a$ conjugate

$P_b + jQ_b = V.I_b$ conjugate

P + jQ = V.I conjugate

Therefore,

$P_a + jQ_a = (P + j\,Q)\times$ (**Z_b conjugate/(Z_a + Z_B)**conjugate)

$P_b + jQ_b = (P + j\,Q)\times$ (**Z_a conjugate/(Z_a + Z_B)**conjugate)

kVA = 600

p.f. = 0.8 lag

P = 480 kW

Q = 360 kVAr

$R_a = 0.008\ \Omega$

$R_b = 0.003\ \Omega$

$X_a = 0.035\ \Omega$

$X_b = 0.019\ \Omega$

$$P_a = \frac{(P.R_b + X_b).(R_a + R_b) - (Q.R_b - P.X_b)(X_a + X_b)}{(R_a + R_b)^2 + (X_a + X_b)^2} = 172.946\ \text{kW} \quad \textbf{Answer}$$

$$R_a = \frac{(Q.R_b - P.X_b) + (P_a + R_b) + (P.R_b + Q.X_b)(X_a + X_b)}{(R_a + R_b)^2 + (X_a + X_b)^2} = 118.103\ \text{kVAr}$$

$P_b = P - P_a$	$= 307.053$	kW	**Answer**
$Q_b = Q - Q_a$	$= 241.897$	kVar	**Answer**
kVA of A	$= 209.426$	kVA	**Answer**
p.f. of A	$= 0.826$	lag	**Answer**
kVA of B	$= 390.891$	kVA	**Answer**
p.f. of B	$= 0.786$	lag	**Answer**

Prob.6.2.5 Transformers- single phase-parallel operation-Example 3

Two transformers A and B operate in parallel on both sides. Their ratings are as follows:

kVA	*No-load voltage*	*Z%*
Transformer A 300	*240V*	*1 + j6*
Transformer B 200	*235V*	*1.2 + j4.8*

Find

(a) the value of the circulating current at no-load

(b) the load supplied by each, and their respective power factors, when the total load is 400 kVA at 0.8 p.f. lag. Assume a load voltage of 230 V

Solution.

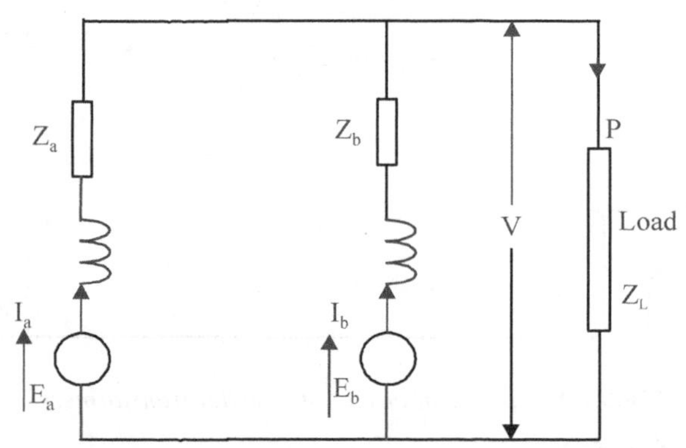

Fig.6.2.5. *Equivalent circuit of parallel transformers.*

$$\mathbf{V = E_a - I_aZ_a = E_b - I_bZ_b = (I_a + I_b)Z_L}$$

$$\mathbf{E_a = E_b + I_aZ_a - I_bZ_b = (I_a + I_b)Z_L + I_aZ_a} \quad \ldots (1)$$

$$\mathbf{E_b = (Ia + Ib)Z_L + I_bZ_b} \quad \ldots (2)$$

$$I_a = \frac{E_a - E_b + I_b.Z_a}{Z_a} \quad \ldots (3)$$

$$I_b = \frac{E_b - E_a + I_aZ_a}{Z_b} \quad \ldots (4)$$

Substituting the values of I_a in (2) and I_b in (1);

$$\mathbf{E_a = [I_a + (E_b - E_a + I_aZ_a) \times \frac{Z_L}{Z_b} + I_aZ_a]} \quad \ldots (5)$$

$$\mathbf{E_b = [I_b + (E_a - E_b + I_bZ_b) \times \frac{Z_L}{Z_b} + I_bZ_b]} \quad \ldots (6)$$

From (5)

$$I_a = \frac{E_aZ_b.-(E_a - E_b).Z_L}{Z_L(Z_a + Z_b) + Z_aZ_b}$$

From (6)

$$I_b = \frac{E_b Z_a . -(E_a - E_b).Z_L}{Z_L(Z_a + Z_b) + Z_a Z_b}$$

kVA common base kVA = 300

V_b base V = 230 Volts

I_b = 1304.347 A

kVA_a = 300

kVA_b = 200

ReZ_a = 0.01 pu

ImZ_a = 0.06 pu

ReZ_a on common base = 0.01

ImZ_a on common base = 0.06

$Z_a$mag = 0.060

Z_a angle = 1.405 rad

ReZ_b = 0.012 pu

ImZ_b = 0.048 pu

ReZ_b on common base = 0.018 pu

ImZ_b on common base = 0.072 pu

Z_b mag = 0.0742

Z_b angle = 1.325

Re $(Z_a + Z_b)$ = 0.022 pu

Im $(Z_a + Z_b)$ = 0.132 pu

$(Z_a + Z_b)$ mag = 0.133

$(Z_a + Z_b)$ angle = 1.405

load kVA = 400

load p.f. = 0.8 lag

P load kW = 320 kW

P = VI cos ϕ = V Cos f V/Zl_{mag}

Therefore ,

$Z_L$mag $= \dfrac{V^2 \cos\phi}{p} = 0.132\ \Omega$

I = V/Zl_{mag} 1739.130 A

I pu = 1.333 at .8 pf lag

Re I = 1.066

Im I = 0.8

I angle = – 36.9° = – 0.6437 rad.

I_a = $IZ_b/(Z_a + Z_b)$

I_a mag = 0.739 pu = 964.507 A **Answer**

I_a angle = -0.723 rad

Cos ϕ_a = 0.749 lag **Answer**

I_b = $IZ_a/(Z_a + Z_b)$

I_b mag = 0.606 pu = 790.513 A **Answer**

I_b angle = – 0.643 rad

Cosϕ b = 0.799 lag **Answer**

$$\text{Circulating current} = \frac{E_a - E_b}{Z_a + Z_b} = \frac{V + I_a Z_a - V - I_b Z_b}{Z_a + Z_b}$$

$$= \frac{I_a Z_a - I_b Z_b}{Z_a + Z_b}$$

Let A $= \mathbf{I_a Z_a / (Z_a + Z_b)}$

A mag = 0.336

A angle = – 0.723

B $= \mathbf{I_b Z_b / (Z_a + Z_b)}$

B mag = 0.336

B angle = 0.299

ReA = 0.251

ImA = – 0.222

ReB = 0.321

ImB = 0.099

Therefore ,

Circulating current = 0.329 pu = 429.207 A

Prob.6.2.6 Transformers -tap changing

Fig.6.2.6 (a) shows a transformer with maximum tap ratio of 0.9 :1. If the maximum overall voltage drop is to be 10 % with the transformer taps fully utilized, calculate the reactive power requirement from the compensator.

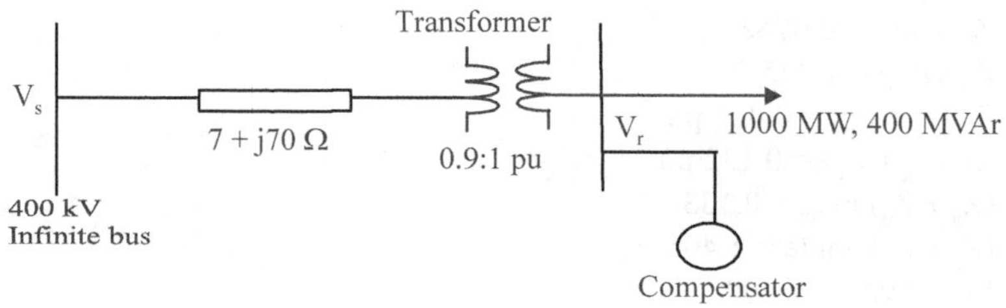

(*a*)

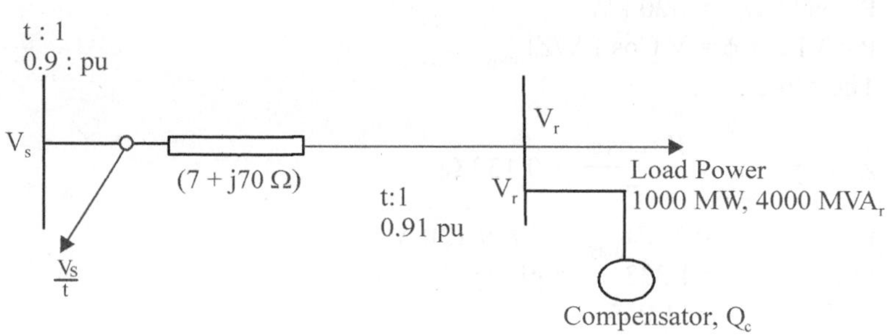

(*b*)

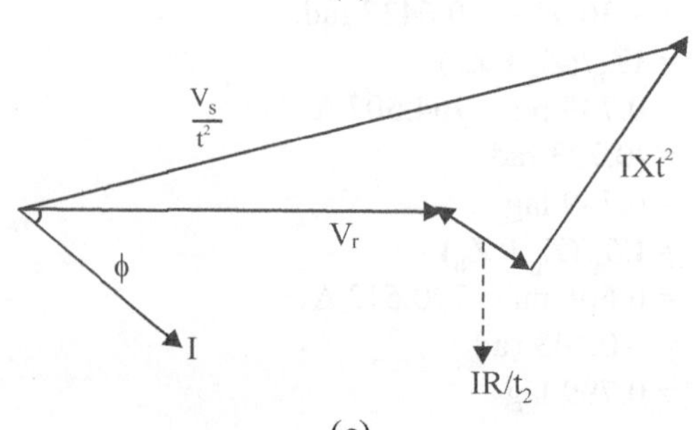

(c)

Fig.6.2.6. *(a) Figure for Prob.6.2.6, (b)(c) Figures for solution of Prob.6.2.6*

Solution. Refer the voltage and line impedance to load side of transformer as shown below

$$\frac{V_s - V_r}{V_r} \times 100 = 10$$

$$\frac{V_r}{V_s} = 0.9$$

$$\frac{V_s}{V_r} = 1.111$$

$$(Vs/t)^2 = A^2 + B^2 \quad \ldots (1)$$

where

$$A = \left(V_r + \frac{I.R}{t^2}.\cos\phi + \frac{I.X}{t^2}.\sin\phi \right) = V_r + \frac{PR}{t^2} + \frac{QX}{t^2}$$

$$B = \frac{IR}{t^2}\sin\phi + \frac{IX}{t^2}\cos\phi = \frac{-RQ}{t^2} + \frac{XP}{t^2}$$

From Eq.1

V_s = 400 kV

V_r = 360 kV

t = 0.9

R = 7 Ω

X = 70 Ω

P = 1000 MW

$$Q = \frac{V_s.V_r.t - R.P - V_r^2 t^2}{X}$$

= 251.771 MVAr

Q_c = MVAr of the compensator

Q_L, Load MVAr = 400 MVAr

$Qc = Q_L - Q = 148.228$ MVAr **Answer**

CHAPTER 7

CIRCUIT-BREAKERS/SUBSTATIONS

7.1 Introduction

Switchgear in a broad sense covers a wide range of equipment connected with switching and protection. A circuit breaker is a switching *i.e.*, current interrupting or making device in switchgear. The basic requirements of switching in power system practice are two-fold :

1. to permit apparatus and circuits to be conveniently put into or taken out of service

2. To permit appropriate and safe isolation of apparatus and circuits automatically, in a pre-determined time period, when they develop faults.

7.1.1 Arc Extinction

Students must be aware of the physical nature of the circuit breaker arc and the three basic methods of arc interruption: high resistance, low resistance (zero point) and vacuum interruption. High –resistance interruption is by lengthening, cooling and/or splitting of the arc, so that its resistance increases until it can no longer maintain itself (used in d.c. and air-break a.c. breakers of low and medium power). Low –resistance interruption aims at keeping the arc energy low by keeping it short and of low resistance and then preventing restriking after a natural current zero; a rapid increase of dielectric strength immediately after a natural current zero is achieved by lengthening and cooling of the contact space and by a 'blast' of oil, air or sulphurhexafluoride (SF_6) across the space (used for medium and high –power circuit-breakers).In the vacuum breaker there are insufficient ion and electrons to maintain the arc, so that rapid extinction with small contact travel occurs(used for low and medium power breakers).

7.1.2 Devices used for circuit breaking (or making)

1. Fuse and Iron clad switches

Fuse is an overcurrent switch in the sense that when the current exceeds a pre-assigned value in a circuit or device, it melts and causes current interruption. The supply is restored only when a healthy one replaces the damaged (melted) fuse in the line. To permit this without any danger of shock to the operator, fuses are connected on the load side of an ironclad switch.

2. Isolators

An isolator is a switch connected after a circuit breaker. When a circuit or a bus bar is taken out of service by tripping the circuit breaker, the isolator is then open circuited and the isolated line is earthed through earth switch so that the trapped line charges are safely conducted to ground.

3. Circuit Breakers

A circuit breaker is a complex circuit -breaking device with the following duties :

- Make or break both normal and abnormal currents.
- Appropriately manage the high-energy arc associated with current interruption. The problem has become more acute due to interconnection of power stations resulting in very high fault levels.

- Effect current interruption only when it is called upon to do so by the relay circuits. In fact they are required to trip for a minimum of the internal fault current and remain inoperative for a maximum of through fault current.
- Rapid and successive automatic breaking and making to aid stable system operation
- 3-pole and single pole auto-reclosing arrangement.

In addition to these making and breaking capabilities, a circuit breaker is required to do so under the following typical conditions :

(*i*) Short-circuit interruption (*i*)
(*ii*) Interruption of small inductive currents
(*iii*) Capacitor switching
(*iv*) Interruption of short-line fault
(*vi*) Asynchronous switching

7.1.3 Classification of circuit breakers

Sketch the constructional details of he following :

1. Air-break circuit breaker or miniature C.B.
2. Oil circuit breaker
3. Minimum oil circuit breaker
4. Air blast circuit breaker
5. SF_6 circuit breaker
6. Vacuum circuit breaker

The main constructional features of the different types of breaker should be understood. The advantages and/or necessity for the modular type of construction for H.V. units should be known. Students should also be aware of the auxiliary equipment required, e.g., isolating and earthing switches, operating mechanisms, oil-purification plant, compressed -air plant and fire fighting equipment, and should also appreciate the problems of locating the current transformers.

- Explain D.C. and A.C. arc interruptions and bring out the difference between the two.
- Describe briefly the methods adopted for rapid increase of the dielectric strength in various types of circuit breakers.
- Explain the working of a vacuum circuit breaker.

7.1.4 Principles of circuit breaking

D.C. circuit breaking effect of decreasing current & increasing arc length.

A.C. circuit breaking - current - zero period - Distortion of AC current wave by arc voltage recovery and restriking voltages: single-frequency & double frequency transients- rate of rise of recovery voltage (RRRV)-control of RRRV-Resistance switching.

Current chopping-Interruption of low magnetizing currents-Opening resistors-capacitive current breaking –

Switching of capacitor banks and unloaded lines – Interrupting terminal faults & short - line faults

- Explain the phenomenon of A.C. circuit breaking and the factors upon which the restriking voltage depends.
- Explain current chopping and effects of breaking capacitive currents. How are these problems met in practice ?

7.1.5 Ratings of circuit breakers

- Rated voltage
- Rated insulation

- Rated frequency
- Rated normal current
- Rated short circuit breaking current
- Rated short circuit making current
- Rated opening sequence for auto-reclose CBs
- Rated transient recovery voltage for terminal faults (Representation of TRV by 4-parameters and 2- parameters CB interrupting time-its components in relation tofault clearing time).

Single-pole auto reclosing & its effects on system performance.

Rating is generally based on symmetrical 3-phase and simple earth-fault current, and students should be able to calculate such currents.

Methods of full-scale (up to 5000 MVA at 275 kV), unit and synthetic testing should be fully understood together with plant and measuring equipment required and the procedure for carrying out such tests.

7.1.6 Substations

Indoor switchgear

(a) Stationary cubicle type (b) Draw-out or truck type
(c) SF_6 filled switchgear (d) Fuse-switch units
(e) Flame proof or explosion proof switchgear
(f) Cellular type (g) Corridor switchboard
(h) Mimic diagram board (i) Metal-clad switchgear
(j) Isolator and earthing switch-vertical break isolator-Double break isolator

- Sketch the layout of a typical sub-station including bus bar arrangements
- Define and discuss - rated short circuit breaking current - rated short - circuit making current-Rated TRV and its influence on the breaking & making current
- Explain the need of an isolator in a switching station

7.1.7 Extracts from Indian Standards for Sub-station Equipments

LEAD ACID STORAGE STATIONARY BATTERIES [IS: 1651-1991]

TYPE TESTS

S. No.	*Type test*	*Purpose*
1.	Verification of Constructional Requirements	To verify materials of container, top cover, connector, & positive & negative element (lead/calcium) of a cell
2.	Verification of Marking	To know the type of cell, make, manufacturing date etc
3.	Verification of dimensions	To verify dimensions as per rated capacity
4.	Test for capacity	To confirm the rated capacity of a cell
5.	Tests for voltage during discharge	To observe cell voltage during AH capacity test as under (a) Voltage after 10 min. load – 1.98 V (b) Voltage after 6 hours load- 1.92 V (c) Voltage after 10 hours load completed- 1,85 V
6.	Ampere-hr and W-hr efficiency test WH capacity shall not be less than 75%	(a) AH capacity shall not be less than 90%b)
7.	Test for loss of capacity on storage	Shall be less than 10%
8.	Endurance test	To apply 13 cycles of 2000 hours (charging & discharging) and to observe capacity of cell after each cycle discharge. Capacity should exceed 90%

ACCEPTANCE TESTS *Type test Nos. 2, 3, 4, and 5 will be used as the Acceptance tests.*

Sample Scheme

No. of cells in a Lot	*No. of cells as sample*			*Permissible defectives*		
	Stage 1	*Stage 2*	*Stage 1 2 + stage*	*Stage 1*	*Stage 2*	*Stage 1 + stage 2*
Up to 50	2	2	4	0	1	1
51-300	3	3	6	0	1	1
301-500	5	5	10	0	2	2
501-1000	8	8	16	0	2	2
1001 & above	13	13	26	0	3	4

NICKLE CADMIUM STORAGE STATIONARY BATTERIES [IS: 10918-1984]

CIRCUIT BREAKERS [IEC: 62271-100]

TYPE TESTS

S. No.	*Type test*	*Purpose*
1.	Dielectric tests	To check characteristics of Circuit breaker for the following tests : I) Dry, wet, power frequency I) Lightning impulse voltage III) Switching impulse voltage (applicable for 420 kV CB and above)
2.	Radio Interference voltage test	To determine RIV on CB pole in both close & open position
3.	Resistance of the main circuit	To record circuit resistance during temp. rise test at 20°C
4.	Temp. rise test	To ensure capability of contacts to carry rated normal current within specified temp. rise limits
5.	Short-time withstand current & peak withstand current	To check the ability of the circuit to carry the maximum rated short-circuit withstand (2.5 times the r.m.s. value) at 50 Hz in close position during the specified short-term duration of 1 or 3 sec.
6.	Mechanical operation test at ambient temp.	To check the characteristics of the breaker for 2000. operations on each pole (with multiple circuit breaker with individual drive) OR complete assembled breaker (mechanically gang-operated breaker with one common drive).
7.	Short-circuit current making/breaking tests	To check ability of the CB to clear the current on different tests in symmetrical & asymmetrical conditions as per provisions of IEC.
8.	Capacitive current charging tests :- – Line charging current breaking tests – Cable charging current breaking tests	To check withstand capability of the CB for no load transient lines, cables, and capacitor banks.

ROUTINE TESTS

S. No.	*Routine test*	*Purpose*
1.	Dielectric test on main circuit	To check the dielectric's withstand capability of live terminals to live terminals and to earth in both close & open condition of CB.
2.	Dielectric test on auxiliary & control circuit	To check the dielectric's withstand capability of auxiliary & control circuit of CB subject to short duration (60 sec) voltage withstand test for 2 kV. (For motor or other devices , they are subjected to a dielectric test as per appropriate specs.)

3.	Design identification test	Verifies the CB for compliances in terms of language of name plate, identification of aux. Equipment, colour & quality of paint etc.
4.	Mechanical operating test	Includes 1. 5 open-close operations at max. rated. and minimum control voltage. 2. 5 close- open operation at rated control voltage 3. recording opening & closing times at the rated operating pressure & voltage.
5.	Measurement of resistance of main circuit	To record contact resistance of CB for mechanical operations to be within specified limits.

THE ROUTINE TESTS ARE TO BE COSIDERED AS ACCEPTANCE TESTS

SAMPLE CRITERION

It is general practice to select samples equal to 10% of offered lot.

Indian Standards for CTs/ PTs

CURRENT TRANSFORMERS [IS: 2705 (Part-1)-1992]

Notes

1. After the CTs have been subjected to dielectric tests (3, 4, and 5), all routine tests will be carried out.
2. Type tests may be omitted when the manufacturer hold certificate of type tests on a similar transformer, which is acceptable to the purchaser.

Routine Tests

S. No.	***Routine test***	***Purpose***
1.	Verification of terminal marking & polarity	To ensure correct marking & function of CT
2.	Power frequency dry withstand test on primary winding	To verify power frequency withstand capacity between primary winding & earth
3.	Power frequency dry withstand test on secondary winding	To verify power frequency withstand capacity between secondary winding & earth
4.	• Over voltage inter-turn test • Partial discharge test • Current error test • Phase displacement error • Accuracy class • Burden	• To ensure inter-turn insulation strength • To ensure insulation strength • To measure actual current ratio against rated ratio • To measure phase displacement between primary & secondary currents • To know the actual accuracy against rated • To measure the impedance of the secondary circuit

OPTIONAL TESTS

S. No.	***Optional test***	***Purpose***
1.	Chopped lightning impulse test as a type test	To verify negative polarity lightning impulse voltage withstand capacity of external insulation.
2.	Measurement of dielectric dissipation factor for oil immersed CT of 72.5 kV & above	To measure dielectric dissipation factor (tan δ) of primary winding between ambient temperature & equipment temperature under test.
3.	Commissioning tests on new CT up to & including 36 kV	To measure withstand capacity of primary winding at 80 % of rated over voltage at site.

ACCEPTANCE TESTS

The routine tests are to be considered as acceptance tests.

SAMPLE CRITERIA

It is general practice to select samples equal to 20 % of offered lot.

VOLTAGE TRANSFORMERS (P.T) [IS: 3156 (Part-1)-1978

The tests are similar to CT tests. For determining the burden, the admittance (mhos) of secondary circuit & the power factor are measured.

Indian Standards for Isolators, Lightning Arrestors, Switchgear Panels, LV Panels, and Gas insulated substations

AC DISCONNECTORS (ISOLATORS) & EARTHING SWITCHES FOR VOLTAGE 1000 V [IS: 9921 (Part-4-1985)]

TYPE TESTS

S. No	*Type test*	*Purpose*
	(A) NORMAL TYPE TESTS	
1	Dielectric tests	To verify insulation level
2	Temperature rise tests	To verify that the temperature rise of any part does not exceed the limit
3	Short-time withstand current & peak withstand current tests	To prove the capability of the switches to withstand the stated currents
4	Short-circuit making performance of earthing switches	To prove the capability of the switches to carry the short-circuit making current
5	Operating & mechanical endurance tests	To prove satisfactory operation & mechanical endurance
(B) TYPE TESTS ON SPECIAL REQUEST OF USER		
6.	Operation under severe ice conditions	To prove satisfactory operation under severe snowfall conditions
7.	Operation at temperature limits	To prove satisfactory operation under minimum/ maximum ambient temperature
8.	Measuring radio interference level	For rated voltage > 120 kV

Note: The entire test shall be carried out on complete isolator or earth switch on their operating devices & auxiliary equipments. Single –pole tests are however permitted in certain cases.

ROUTINE TESTS

S. No	***Routine test***	***Purpose***
1.	Power frequency voltage withstand test of main circuit.	To verify insulation level including withstand tests of power frequency voltages on auxiliary equipment.
2.	Voltage test on control & auxiliary circuit.	To verify insulation level.
3.	Resistance measurement of main circuit.	Should not exceed type test value.
4.	Mechanical operation tests.	To ensure that disconnectors/earthing switches . comply with operating conditions at rated, maximum, and minimum supply voltage.

Note: Where isolators/earthing switches are not assembled at the manufacturer's works to enable the routine tests to be made, tests on part assemblies should be subjected to under agreement between the manufacturer & the user.

The routine tests are to be considered as *ACCEPTANCE TESTS*

SAMPLE CRITERIA: It is general practice to select samples equal to 10% of offered lot.

METAL OXIDE SURGE ARRESTORS WITHOUT GAP FOR AC SYSTEM [IEC 60099-4 & IS: 3070 –1993]

TYPE TESTS [as per IEC]

S.No	*Type test*	*Purpose*
1.	Insulation withstand test	Demonstrates the ability of the housing to withstand voltage stresses during dry /wet conditions.
2.	Residual voltage test	Demonstrates the protective levels of the arrestor.
3.	Long duration current impulse test	Demonstrates the ability of he resistor elements to withstand dielectric & energy stresses without puncture or flashover.
4.	Operating duty test	Defines the thermal stability of the arrestor under defined conditions.
5.	Pressure relief test	Demonstrates the ability of the arrestor housing to withstand short-circuit currents without violent shattering of the housing under test conditions.
6.	Test of arrestor disconnectors	Demonstrates correct operation of the disconnectors
	Artificial pollution test (not yet established)	To show that the internal parts of he arrestor including grading system are able to withstand pollution without any damage and that the external insulation does not flashover.
7.	Partial discharge test.	Measures the internal partial discharges
	Seal leakage test (not yet established)	Determines the integrity of the arrestor seals
8.	Current distribution test for multi-column arrestor.	Determines the current through each column of parallel resistor when there is no electrical connection between the columns.

Routine Tests

S. No	*Routine test*	*Purpose*
1.	Measurement of reference voltage	Measured values to be within a range specified by the manufacturer
2.	Measurement of power frequency voltage on complete arrester	Measured values to be within a range specified by the manufacturer
3.	Lightning impulse residual voltage test	Measured values to be within a range specified by the manufacturer

NOTE: Type tests 2. 7 and 8 will be taken as routine tests

SPECIAL TEST

S. No	Special test	Purpose
1.	Special thermal stability test	Metal oxide arrester temperature or resistive component of current or power dissipation be monitored during power frequency voltage application to prove thermal stability.

SAMPLE CRITERIA

[As per clause No. 8.2.1 of IEC: 60099-4:1998]
The acceptance test shall be made on the nearest lower whole number in the cube root of the number of arrestors to be supplied.

Indian Standards for LV Panel Boards [IS: 8623-1993]

There is no specific Indian Standard for the tests to be performed. However the manufacturer shall carry out following tests on the low voltage panel board and ask for type/routine tests of bought out items like switch fuse units, meter, fuses, copper strip, indicating instruments etc.

Ref.: [IS 8623-1993]

TYPE TEST

S. No.	Type test	Purpose
1.	Temperature rise test	To verify the temperature rise of IT panel board including all bus bars and equipments like MCCB, ACB, MCB, Switch disconnectors.
2.	Dielectric properties test	To verify withstand capacity in HV test.
3.	Verification of the effective connection between the exposed conductive parts of the ASSEMBLY and the protective circuit the inspection or by resistance measurement	To verify connection between the exposed conductive parts and the protective circuit.
4.	Verification of clearances and creepage distances	To check for minimum required clearances.
5.	Verification of mechanical operation	To verify that product is suitable for N number of operations without any problem.
6.	Verification of degree of protection	To ensure protection against ingress of material as per the various IP code.
7.	Inspection of Assembly including inspection of wiring and, if necessary, electrical operation test	To check wiring and its proper connections.
8.	Dielectric test	To verify insulation resistance.
9.	Checking of protective measures and the electrical continuity of the protective circuits	To check protective measures.
10.	Insulation resistance test	To verify insulation provided on conductive parts and parts with the body.
11.	Verification of Name plate /rating plate of all inserts	To confirm use of all inserts as per approved drawing/ bill of materials.

Notes:

- The routine tests are to be considered as acceptance tests.
  The routine test certificates of manufacturer of all bought out items are to be verified.

SAMPLING CRITERIA

Select samples equal to 10 % of offered lot. Alternatively, as per sampling criteria given in IS: **2500**

SWITCHGEAR PANELS WITH HV A.C. CIRCUIT BREAKER [IS: 13118-1991]

TYPE TESTS

S.No	Type test	Purpose
1.	Dielectric tests: • 1.1 Lightning & switching impulse voltage test • 1.2 Power frequency voltage test • 1.3 Artificial pollution test • 1.4 Partial discharge test • 1.5 Tests on auxiliary & control circuits	To verify that complete circuit breaker panel along with auxiliary and control circuits can withstand relevant dielectric tests
2.	Radio interference voltage test (on one pole of CB during open & closed position	To check RIV during high voltage.
3.	Temperature rise test	To record actual temperature rise of breaker panel including all bus bars to be within specified limits.
4.	Measurements of contact pressure & resistance of main circuit	To verify the contact pressure and resistance of main circuit as per designed value.

5.	Short-time withstand & peak withstand test	To verify the short-circuit withstand capacity capacity of C.B Panel
6.	Mechanical & Environment test	
	Mechanical operation test at ambient air temperature	To confirm that the product is suitable for N number of operations without any problem.
	Low & high temperature test	To verify temperature rise of CB & its mechanism to be within specified limit.
	Humidity test	To check withstanding CB against humidity
7.	Making & Breaking tests -Single phase making capacity test -Single phase breaking capacity test	To verify single phase making/braking capacity of CB
8.	Basic short-circuit tests- Duties Nos.1-5 tests	To verify withstand capacity of CB for various Short-circuit Duties.
9.	Critical current test	To verify withstanding critical current capacity of CB. Applicable only to CB which have critical current of less than 10% of rated SC breaking current.
10.	Single phase SC test	Applicable to three pole CB To check adverse effect in operation of CB due unbalanced force produced.
11.	Short line fault test	Applicable to 3 pole CB directly connected to overhead lines. To verify withstanding capacity of CB for various SC duties.
12.	Out of phase making/breaking test	To determine ability of CB to break & make currents during out of phase conditions.
13.	Capacitive current switching tests -Rated line charging breaking current -Rated cable charging breaking current -Rated single capacitor bank breaking current -Rated back to back capacitor bank breaking current -Rated capacitor bank inrush making current	To check withstanding capacity of CB during line/cable/ capacitor bank charging or breaking current.
14.	Small inductive current switching test	To check withstanding capacity of CB during inductive current.

ROUTINE TESTS

S. No	*Routine test*	*Purpose*
1.	Dielectric tests: • Dry power frequency test • Voltage withstand tests on auxiliary & control circuits	To verify that complete circuit breaker panel along with auxiliary and control circuits can withstand relevant dielectric tests.
2.	Mechanical operation test at ambient air temperature	To confirm that the product is suitable for N number of operations without any problem.
3.	Visual & design check	
4.	Mechanical operation test at ambient air temperature	To confirm that the product is suitable for N number of operations without any problem.

Notes.

- The routine tests are to be considered as acceptance tests.
- The routine test certificates of manufacturer of all bought out items are to be verified.

SAMPLING CRITERIA

Select samples equal to 10 % of offered lot.

Indian standards for Gas- insulated substations [IEC: 517 and IEC-694]

OBJECTIVE TYPE QUESTIONS

1. Which of the following results when a resistance is placed across the contacts of a circuit breaker ?
   (*i*) successful and effective interruption of current
   (*ii*) undue power loss in the resistor
   (*iii*) effective fault power dissipation
   (*iv*) control of rate of rise of restriking voltage.
   (*a*) all four above
   (*b*) (*i*) and (*iii*)
   (*c*) (*i*) and (*iv*)
   Ans. (*c*)
2. The circuit breaker rating relevant to transient stability is :
   (*a*) rated voltage
   (*b*) rated operating time
   (*c*) rated breaking current
   Ans. (*c*)

PROBLEMS WITH SOLUTIONS

Problem 7.2.1

Fig. 7.2.1 shows a 6.6 kV, 10 MVA generator with a reactance of 0.12 pu. Calculate the fault current for a fault close to the bus bars.

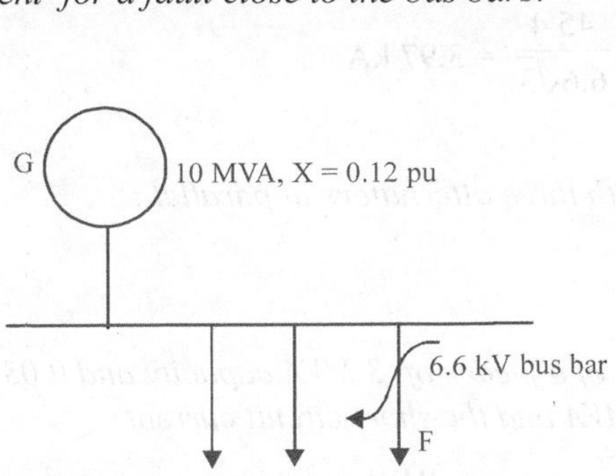

Fig.7.2.1. *Fault close to the bus bar*

Solution.

$$\text{Normal Current} = \frac{10.000 \times 1000}{\sqrt{3 \times 6600}} = 878 \text{ A}$$

Shirt-circuit MVA = 10/0.12 = 83.3 MVA **Answer**

$$\text{Short-Circuit Current} = \frac{83.3}{\sqrt{3 \times 6600}} = 7.28 \text{ kA}$$ **Answer**

This is the initial value of the short-circuit current; if the interruption were delayed by a few cycles, the generator decrement would reduce, by an appreciable amount the current that actually had to be interrupted by the circuit-breaker. If, however, the circuit-breaker were to be closed on to the short-circuit, the initial peak-making current would be $7.28 \times 1.8 \times \sqrt{2} = 18.6$ kA. The circuit-breaker would have to be designed to withstand the electromagnetic forces due to this current.

Problem 7.2.2

Find the short-circuit MVA and the short-circuit current for the same system as in Prob.7.2.1, but with the fault F at the far end of one of the feeders.

The full-load capacity of the feeder is 5 MVA ,and its reactance is 0.05 pu.

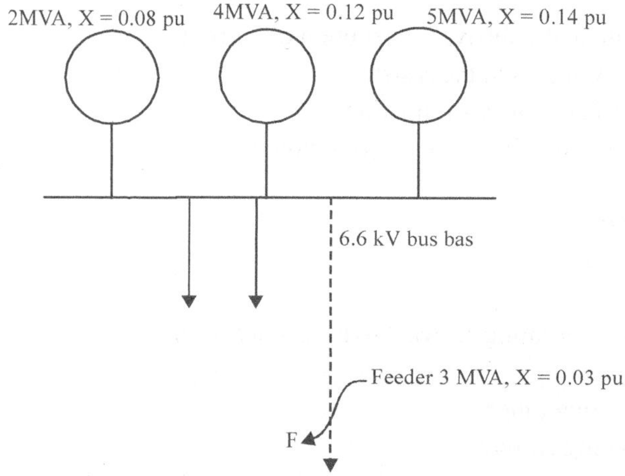

Fig.7.2.2. *Fault at the far end of the feeder*

Solution.

Feeder reactance based on 10 MVA = $\frac{10}{5}\times 0.05$ = 0.1 pu

Total reactance based on 10 MVA = 0.12 (generator) + 0.1(feeder) = 0.22 pu

Short-circuit MVA = 10/0.22 = 45.4 MVA **Answer**

Short-circuit current = $\frac{45.4}{6.6\sqrt{3}}$ = 3.97 kA **Answer**

Problem 7.2.3

Fig. 7.2.3 shows a system with three alternators in parallel :

A 2 MVA, 0.08 pu reactance

B 4 MVA, 0.12 pu reactance

C 5 MVA, 0.14 pu reactance

A fault occurs at the far end of a feeder of 3 MVA capacity and 0.03 pu reactance. Calculate the short-circuit MVA and the short-circuit current.

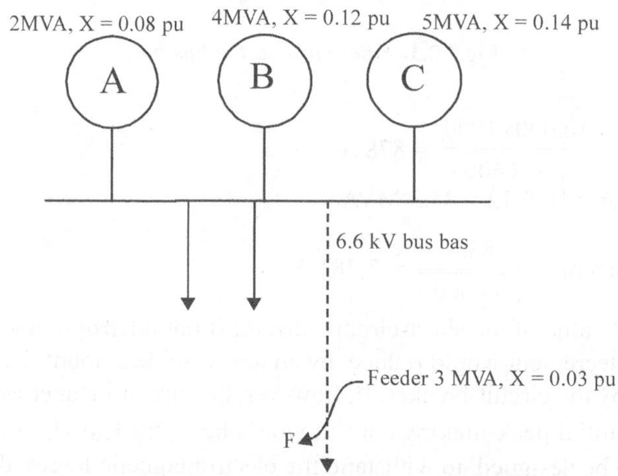

Fig.7.2.3. *Fault at the far end of the feeder*

Solution.

Total generator capacity = 2 + 4 + 5 = 11 MVA

Normal current based on 11MVA $\frac{11}{6.6\sqrt{3}}$ =0.965 kA

Reactance of generator A based on 11 MVA= $\frac{11\times0.08}{2}$ =0.44pu

Reactance of generator B based on 11 MVA= $\frac{11\times0.08}{4}$ =0.33pu

Reactance of generator based on 11 MVA = $\frac{11\times0.14}{5}$ = 0.31pu

These three reactances are in parallel. The total machine reactance to the

same base is thus = $\frac{1}{\left(\frac{1}{0.44}\right)+\left(\frac{1}{0.33}\right)+\left(\frac{1}{0.31}\right)}$ = 0.117 pu

Feeder reactance based on 11 MVA = 11*0.03/3 = 0.11 pu
Total reactance up to fault = 0.117+0.11= 0.227 pu
Short-circuit MVA = 11/0.227 = 48.3 MVA **Answer**

Short-circuit current = $\frac{48.3}{6.6\sqrt{3}}$ = 4.23 kA **Answer**

Problem 7.2.4

The system shown in Fig.7.2.4 comprises three generators, two step-up transformers, two transmission lines in parallel, two step-down transformers, and secondary feeders each with step-down transformers supplying distributers and a fault on one of the distributors near the transformer.

The three generators have the same capacities and reactances as in Prob.7.2.3.
Calculate the short-circuit MVA and the short-circuit current.

Solution.

The normal current based on machine capacity of 11 MVA = 0.965 kA
Reactance of alternators based on 11 MVA = 0.117 pu

Reactance of step-up transformers based on the same MVA = $\frac{11\times066}{5}$ = 0.132 pu

Reactance of line to same base = $\frac{11\times005}{5}$ = 0.11pu

Reactance of step-down transformer o same base = $\frac{11\times006}{5}$ = 0.132 pu

Total reactance of one line with transformers = 0.132 + 0.11+0.132 = 0.37 pu
Total reactance of two lines in parallel = 0.185 pu

Reactance of secondary line to 11 MVA base = $\frac{11\times0.04}{1}$ = 0.44 pu

Reactance of 1 MVA transformer to same base = $\frac{11\times0.03}{1}$ =0.33 pu

Total reactance up to the fault point = 0.117 + 0.187 + 0.44 + 0.33 = 1.07 pu
Short-circuit MVA = 11/1.07 = 10.22 MVA

$$\text{Short-circuit current} = \frac{10.28}{0.5\sqrt{3}} = 11.87\,\text{kA}$$

With a fault so far away from the generators, the generator reactance is only small proportion of the total and the generator decrement of current will only have a negligible effect. It is clear from this example that lines or transformers of small capacity present a special feature ,since their reactance when expressed on the basis of the total plant capacity is very high.

Thus ,suppose one of the distributors had a transformer of capacity 100 kVA and reactance = 0.05 pu, then this transformer could not pass more than100/0.05 = 2000 kVA = 2MVA, no matter how large the total capacity of the generating plant might be.

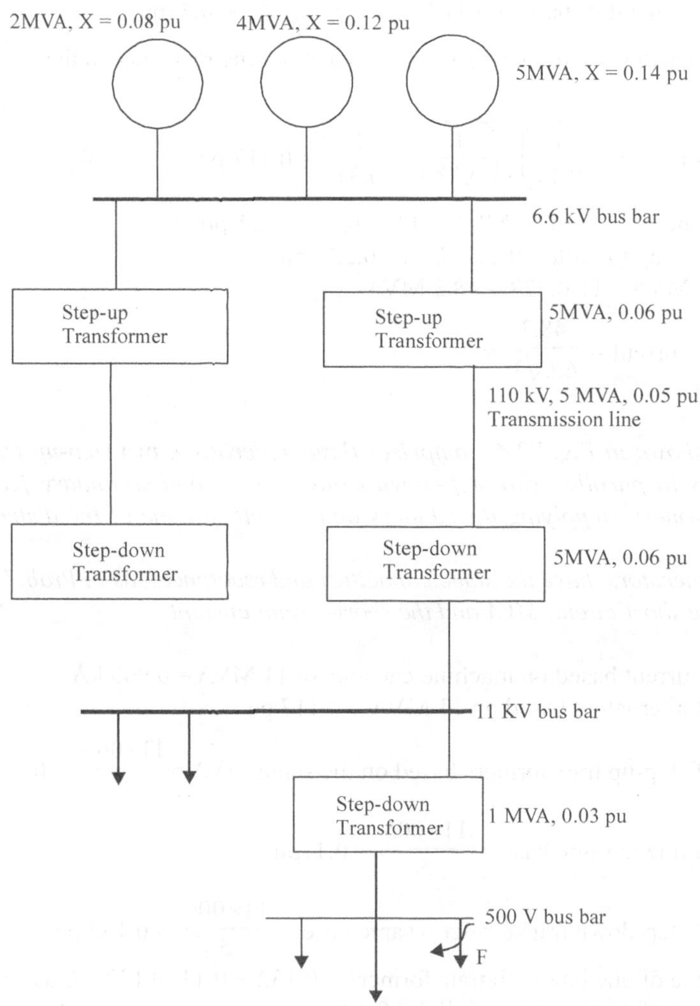

Fig.7.2.4. *A power system with a fault on the distribution side.*

CHAPTER 8

LOAD FLOWS

8.1 INTRODUCTION

8.1.1 General introduction & importance of load flows

- Concept of power balance at each node in the power network. Real & reactive power flows. State vector and utility of the information contained in the state vector for various studies (such as planning & operation) in the power system.
- Buses, injected power at each bus, real & reactive power balance equations. Properties & formulation of Y bus. Formulation of Y bus by observation. The impedance matrix.
- Classification of buses. Specification of variables at different buses. Definition & need of slack bus. Physical interpretation of the power flow equations. The nature of power flow equations.

8.1.2 Solution of load flow equations

- Emphasise the difficulty in obtaining a closed form solution of the power flow equations. The need for iterative solution methods. Gauss iterative method. Steps in the iterative solution. Convergence criterion. Acceleration factors.
- Gauss-Seidel method of solution. Steps in the iterative solution. Introduction to the Newton -Raphson method.
- Problems range from simple calculations of active and reactive power flow in two parallel circuits (including the effect of transformer tap-changing and quadrature boosters) up to power distribution over a large network. Students should fully understand the setting up of the node and mesh equations. Direct and iterative methods of solution of the equations should be understood, noting that, in practice, node equations are normally used although their numerical solution requires a higher arithmetical accuracy than for the mesh equations.

8.1.3 Websites

Look up the following websites:

• *http://www.geocities.com/SiliconValley/Lab/4223* for information on Sparsity Technology for power system

• *http://www.waterloo.ca/~Tecl:claudio/software/pflow.html* for the software PFLOW: a "commercial-grade" continuation power flow program, that can be used to research static voltage stability phenomena in detailed AC-HVDC-FACTS power system models. The program is written in C and C++, and two versions for WINDOWS 95 and UNIX are available, including examples and a brief tutorial.

• *http://www.ece.utexas.edu/~grady/PCFLO.html* for the softwares PCFLO and PCFLOH PCFLO is a load flow program which also includes short circuit and harmonics capabilities. PCFLOH is a harmonics-only version of PCFLO, with special user interface for harmonic studies. PCFLO contains some features not found in many load flow programs. For example, it lists the power loss on each line and transformer. It allows the user to control the transition point from Gauss-Seidel to Newton-Raphson. Also, it lists the optimal bus order for selected method 1, 2, or 3 in a separate output file, which is helpful to students who are learning how to properly order busses. (for Windows 95/98/NT)

- *http//www.powerworld.com/* for the software PowerWorld™ Simulator: (12 bus version) solves the power flow and economic dispatch problems. It is an interactive power system simulation package designed to simulate high voltage power system operation on a time frame ranging from several minutes to several days. Powerful visualization techniques are used on an interactive basis, resulting in an extremely intuitive and easy to use graphical user interface (GUI). The GUI includes animated one-line diagrams with support for panning, zooming and conditional display of objects. The software runs under Microsoft Windows 95/98/NT.

- *http://ece.uwaterloo.ca/~madriga/freesoft.html* for the software Matlab Power Flow Program

- http://*www.InteractWithPower.com* for the free software (for testing and educational purposes) which is a full AC power flow version. Only standard internet browser needed. No additional software installation needed at your own computer. Steps: Click on "Create account ..."; Specify your data including your correct Email (mandatory) address, accept usage conditions, then choose AC power flow (ready for various "standard" IEEE power flow networks (14, 57,118, 300 IEEE bus systems); you can also feed your own IEEE cdf data sets!). A great opportunity for academics (and professionals) to share great power applications with the power system community.

OBJECTIVE TYPE QUESTIONS

1. Fig.8.2.1, in Problem 8.2.1 below, represents a sample power network. The impedance of the line is marked in the figure, in Ω. The Y– bus (2,2) entry is
(*a*) –j 0.2
(*b*) j 0.2
(*c*) j 5
Ans. (a)

2. The real or reactive power balance equations in a power system network consist of
(*a*) Linear algebraic equations
(*b*) Non-linear algebraic equations
(*c*) Lin ear differential equations
Ans. (b)

3. The presence of phase-shifting transformers in a power system will affect its load flow solution in the following way :
(*a*) Only the diagonal elements of Y – bus will change
(*b*) Y – bus will not be symmetrical
(*c*) Load flow solution cannot be obtained.

PROBLEMS WITH SOLUTIONS

Prob.8.2.1 Two bus power system

Two machines 1 and 2 are connected through an impedance Z as shown:

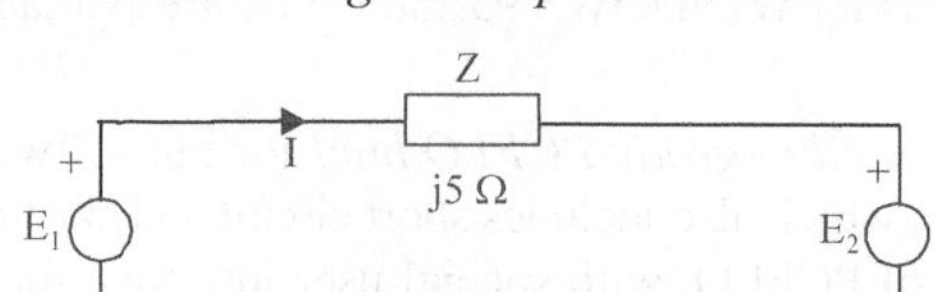

Fig. 8.2.1. *Two bus power system*

$E_1 = 100\angle 0°$ (Volts), $E_2 = 100\ \angle 30°$ (volts), and $Z = j\ 5\ \Omega$.
Determine

(a) the power generated or consumed by each machine
(b) the reactive power received or supplied by each machine
(c) the power and reactive power absorbed by the impedance.

Solution.

$$E_1 = 100$$
$$ImE_1 = 0$$
$$ReE_1 = 100$$
$$E_2 = 100$$
$$E_2 = E_2\cos 30 + jE_2\sin 30$$
$$ReE_2 = 86.615$$
$$ImE_2 = 49.977$$
$$Im\,Z = 5$$
$$\mathbf{I} = (\mathbf{E_1} - \mathbf{E_2})/\mathbf{Z}$$
$$Re\mathbf{I} = (ImE_1\text{-}ImE_2)/Im\mathbf{Z} = -9.995$$
$$Im\mathbf{I} = (Re\backslash E_2\text{-}ReE_1)/Im\mathbf{Z} = 2.676$$

Find E_1.**Iconjugate**

$Re\,(E_1.\mathbf{I}conjugate) = ReE_1.Re\mathbf{I} + ImE_1.Im\mathbf{I} = -999.54$ W **Answer**

$Im\,(E_1.\mathbf{I}conjugate) = Re\mathbf{I}.ImE_1\text{-}ReE_1.Im\mathbf{I} = 267.683$ Var **Answer**

Find $\mathbf{E_2}$.**Iconjugate**

$Re\,(E_2.\mathbf{I}conjugate) = ReE_2.Re\mathbf{I} + ImE_2.Im\mathbf{I} = -999.54$ W **Answer**

$Im\,(E_2.\mathbf{I}conjugate) = Re\mathbf{I}.ImE_2\text{-}ReE_2.Im\mathbf{I} = -267.684$ Vars **Answer**

$$I^2X = \text{reactive power lost in line} = I^2.\mathbf{I_m Z}$$
$$I = 10.347 \text{ A}$$

Reactive power loss in line $= I^2 \times Im\mathbf{Z} = 535.367$ Vars **Answer**

Machine 1 is expected to be a generator because of current direction and polarity.

Machine 1 is actually a motor, consumes 1000W and supplies reactive power 268 vars.

Machine 2 expected to be a motor is actually a generator, generates 1000 W and supplies 268 Vars.

Total reactive power supplied = 2 × 268 = 536 Vars. This is absorbed by the line reactance.

Prob.8.2.2 Nodal voltage equations

Write down the general nodal voltage equations I = Y.V for three nodes and from the data given below determine the value of each of the following complex quantities in pu based on 100 MVA: I_3, Y_{13}, Y_{11}.

Three identical power stations 1, 2 and 3 each contain four identical generator-transformer sets. Each generator has a synchronous reactance of 1.9 pu, and each transformer has a leakage reactance of 0.1 pu. The open-circuit source emfs of the stations are respectively, 2.1∠20°, 2.05 ∠10°, 2 ∠0°.

The stations are interconnected by a closed ring of 275 kV lines having lengths

Node 1 to Node 2 —— 100 km
Node 2 to Node 3 —— 100 km
Node 1 to Node 3 —— 200 km
The line reactance is 0.757 Ω / km

The only load is connected to station 3. All pu data are based on 100 MVA.

Solution. Reactance (generator+Transformer) = 2pu

X_{gt1} Reactance for 4 sets at bus 1 = 0.5pu

X_{gt2}, at bus 2 = 0.5 pu

X_{gt3}, at bus 3 = 0.5 pu

X Line1-2 = 75.7 Ω

Line2-3 = 75.7 Ω

$$\text{Line1-3} = 151.4\ \Omega$$
$$V_b \text{ base value} = 275 \text{ kV}$$
$$V_b.I_b \text{ base value, 3 phase} = 100 \text{ MVA}$$
$$X_{base} = V_b^2/V_b.I_b = 756.25\ \Omega$$
$$X \text{ Line 1-2 pu} = X/X_{base} = 0.10$$
$$\text{Line 2-3 pu} = 0.10$$
$$\text{Line1-3 pu} = 0.20$$
$$\mathbf{E_1} \textbf{ e.m.f.} = 2.1\ \angle 20°$$
$$\mathbf{E_2} \textbf{ e.m.f.} = 2.05\angle 10°$$
$$\mathbf{E_3} \textbf{ e.m.f.} = 2\ \angle 0°$$

We now convert e.m.f.s into current sources and reactances into susceptances to build the nodal admittance matrix.

$$\mathbf{I_1} \text{ current source} = \mathbf{E_1}/jX_{gt1} = 4.2\ \angle -70°$$
$$\mathbf{I_2} \text{ current source} = \mathbf{E_2}/jX_{gt2} = 4.1\ \angle -80°$$
$$\mathbf{I_3} \text{ current source} = \mathbf{E_3}/jX_{gt3} = 4\ \angle -90°$$

Susceptances:

$$b_{gt1} = 1/X_{gt1} = 2 \text{ pu}$$
$$b_{gt2} = 1/X_{gt2} = 2 \text{ pu}$$
$$b_{gt3} = 1/X_{gt3} = 2 \text{ pu}$$
$$b \quad \text{Line1-2} = 9.99 \text{ pu } (= 75.7/756.25)$$
$$\text{Line2-3} = 9.99 \text{ pu}$$
$$\text{Line1-3} = 5.00 \text{ pu}$$

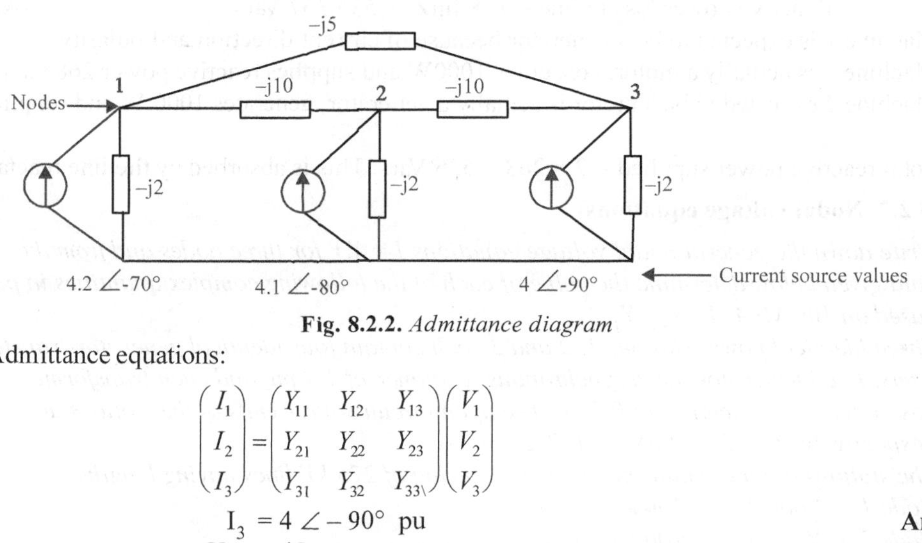

Fig. 8.2.2. *Admittance diagram*

Admittance equations:

$$\begin{pmatrix} I_1 \\ I_2 \\ I_3 \end{pmatrix} = \begin{pmatrix} Y_{11} & Y_{12} & Y_{13} \\ Y_{21} & Y_{22} & Y_{23} \\ Y_{31} & Y_{32} & Y_{33} \end{pmatrix} \begin{pmatrix} V_1 \\ V_2 \\ V_3 \end{pmatrix}$$

$I_3 = 4\ \angle -90° \text{ pu}$ **Answer**

$Y_{13} = j5 \text{ pu}$ **Answer**

$Y_{11} = -\ j10 - j5 - j2 = -\ j17\text{pu}$ **Answer**

Prob. 8.2.3 Network Analysis Using Network Equations

A power system is represented by the equivalent single-phase circuit shown in Fig.8.2.3 (a) below.

Calculate the current in amperes in branch en using the nodal voltage method. The figures shown in the circuit are branch reactances and generator voltages. Assume that the pu quantities are based on 66 kV and 100 MVA.

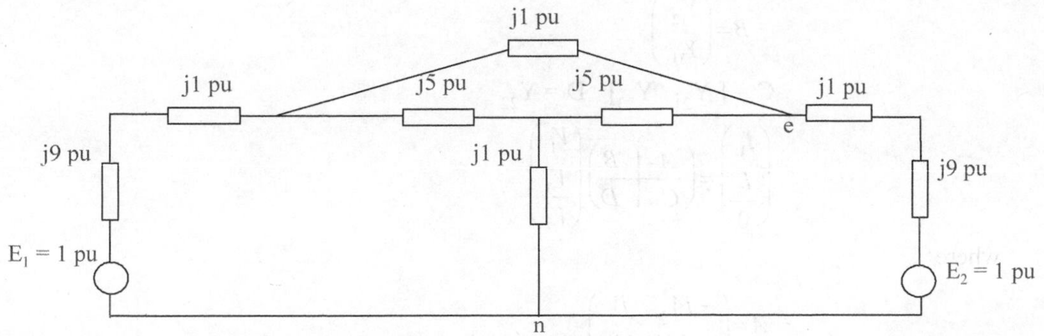

Fig. 8.2.3. (*a*) *Impedance diagram*

Solution.

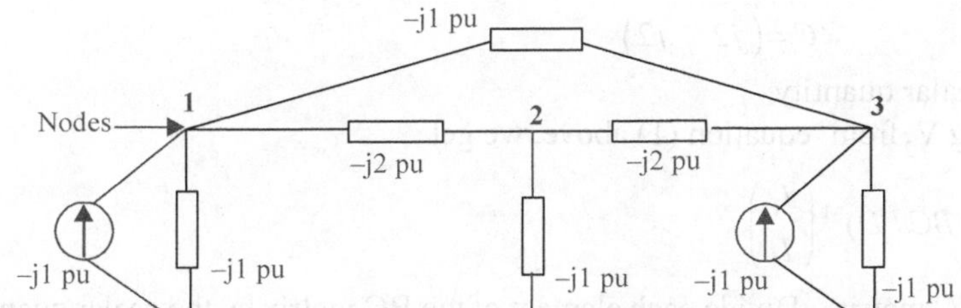

Fig. 8.2.3. (*b*) *Admittance diagram*

Admittance equations :

$$\begin{pmatrix} I_1 \\ I_2 \\ I_3 \end{pmatrix} = \begin{pmatrix} Y_{11} & Y_{12} & L_{13} \\ Y_{21} & Y_{22} & Y_{23} \\ Y_{31} & Y_{32} & Y_{33} \end{pmatrix} \begin{pmatrix} V_1 \\ V_2 \\ V_3 \end{pmatrix}$$

$Y_{11} = -j4$
$Y_{13} = j1$
$Y_{23} = j2$
$Y_{12} = j2$
$Y_{22} = -j5$
$Y_{33} = -j4$
$I_1 = -j1$
$I_2 = 0$
$I_3 = -j1$

Since $I_2 = 0$,we write the following equations :

$$\begin{pmatrix} I_1 \\ I_3 \\ 0 \end{pmatrix} = \begin{pmatrix} Y_{11} & Y_{13} & Y_{12} \\ Y_{31} & Y_{32} & Y_{23} \\ Y_{31} & Y_{32} & Y_{33} \end{pmatrix} \begin{pmatrix} V_1 \\ V_3 \\ V_3 \end{pmatrix}$$

We partition the matrix as above, and let the partitioned submatrices be

$$A = \begin{pmatrix} Y_{11} & Y_{13} \\ Y_{31} & Y_{33} \end{pmatrix}$$

$$B=\begin{pmatrix} Y_{12} \\ Y_{31} \end{pmatrix}$$

$$C=[\,Y_{21} \quad Y_{23}] \quad D=Y_{22}$$

$$\begin{pmatrix} I_1 \\ I_3 \\ \hline 0 \end{pmatrix}=\left(\begin{array}{c|c} A & B \\ \hline C & D \end{array}\right)\begin{pmatrix} V_1 \\ V_3 \\ \hline V_2 \end{pmatrix}$$

where

$$A=\begin{pmatrix} -j4, & j1 \\ j1, & -j4 \end{pmatrix}$$

$$B=\begin{pmatrix} j2 \\ j2 \end{pmatrix}$$

$$C=\begin{pmatrix} j2 & j2 \end{pmatrix}$$

D = – j5, scalar quantity

Eliminating V_2 from equation (1) above, we get

$$\begin{pmatrix} V_1 \\ V_3 \end{pmatrix}=(A-BC/D)^{-1}\begin{pmatrix} I_1 \\ I_3 \end{pmatrix}$$

Calculate BC matrix. Divide each element of the BC matrix by the scalar quantity D, and thus obtain the BC/D matrix.

Then calculate the A – BC/D matrix. The numerical values of A-BC/D matrix are:

$$A-BC/D=\begin{pmatrix} -3.2 & 0.2 \\ 0.2 & -3.2 \end{pmatrix}$$

determinant (A – BC/D) = 10.2

$$(A-BC/D)^{-1}=\begin{pmatrix} 0.313 & -0.019 \\ -0.019 & 0.313 \end{pmatrix}$$

$$\begin{pmatrix} V_1 \\ V_3 \end{pmatrix}=(A-BC/D)^{-1}\begin{pmatrix} I_1 \\ I_3 \end{pmatrix}$$

$$=(A-BC/D)^{-1}\begin{pmatrix} 1 \\ 1 \end{pmatrix}$$

$$=\begin{pmatrix} 0.3333 \text{ pu} \\ -0.333 \text{ pu} \end{pmatrix}$$

base MVA, 3 phase = 100

base kV, (L – L) = 66

I_{base} = 875.810 A

Branch **en** , reactance = 1 pu

Current through branch **en** = $\dfrac{V_3}{\text{branch reactance}}$ = – 0.333 pu

= – 291.93 A **Answer**

Prob.8.2.4 Power system : single-phase equivalent network

A 100 MVA round-rotor generator of synchronous reactance 1.5 pu supplies a substation (L) through a step-up transformer of reactance 0.1 pu (a) two lines each of reactance 0.3 pu in parallel and a step-down transformer of reactance 0.1 pu (Fig.8.2.4.a). The load taken at L is 100 MW at 0.85 pf lagging. L is connected to a local generating station of 75 MVA, and a synchronous reactance of 2 pu. All reactances are on a base of 100 MVA.

Draw the single-phase equivalent network. If the voltage at L is to be 1 pu and the 75 MVA machine is to deliver 50 MW, 20 MVAr, calculate the internal voltages of the machines.

Solution.

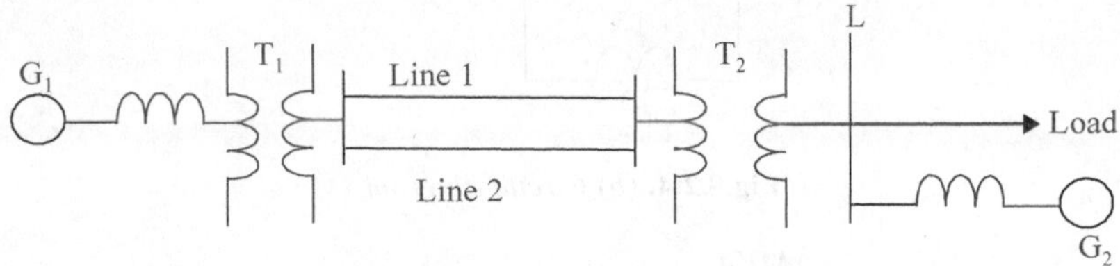

Fig.8.2.4. (*a*) *System diagram*

Base MVA = 100

All reactances are on a base of 100 MVA

G_1:

pu MVA = 1

X = 1.5 pu

T_1

X = 0.1 pu

Line 1

X = 0.3 pu

Line 2

X = 0.3 pu

X of two lines in parallel = 0.15 pu

T_2

X = 0.1 pu

Load

MW = 100

p.f. = 0.85 lag

load MVA = 117.647

load MVA = 1.176 pu

pu MW = 1 pu

pu MVAr = 0.619

G_2:

pu MVA = 0.75 pu

pu MW = 0.5 pu

pu MVAr = 0.2 pu

X = 2 pu

Single-phase equivalent circuit:

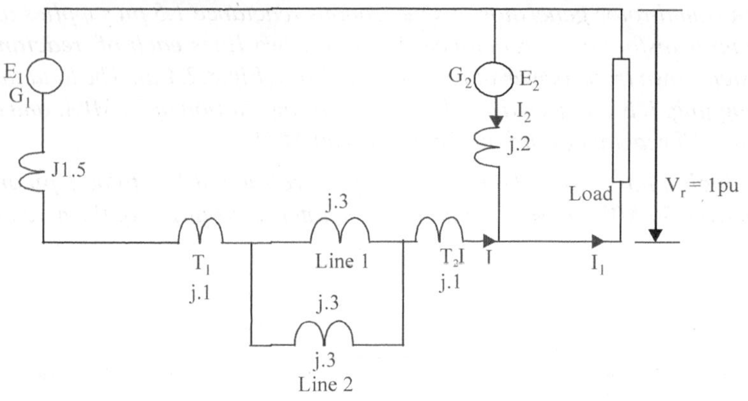

Fig.8.2.4. (*b*) *Circuit diagram*

$$I_1 = \frac{MVA}{1.73 \times V_r \times p.f} = \frac{a}{V_r}, \text{ say}$$

where a = 68.004

I_1 = 68/V_r μA

$$I_{base} = \frac{100}{1.73 \times V_r} = \frac{57.8}{V_r}, \ \mu A$$

I_1 pu = 1.176

$$I_2 = \frac{50}{1.73 \times V_r \times \cos\left[\tan^{-1}\left(\frac{20}{50}\right)\right]} = b/V_r, \text{ say,}$$

where

b = 31.128

I_2 = 31.13/V_r μA

I_2 pu = 0.538 pu

I pu = $I_1 - I_2$ = 0.637 pu

$\mathbf{E_2}$ = $V_r + I_2.j_2$, V_r is referene vector

$\mathbf{E_1 = Z.I + V_r}$

Z = j1.85 (= j1.5 + j0.1 + j0.15 + j0.1)

$\mathbf{E_1 = j1.85\,(I_1 - I_2) + V_r}$

$\mathbf{I_1}$ = I_1mag ∠– Φ

I_1mag = 1.176 pu

$$\Phi = \tan^{-1}\frac{Q_{Load}}{P_{Load}}$$

= – 31.7883°.

$\mathbf{I_2}$

I_2mag = 0.538 pu

$$I_2\text{angle} = \tan^{-1}\left(\frac{Qg_2}{Pg_2}\right) = -21.801°$$

$E_1 = j1.85(1.18.\cos 31.79 - j1.18.\sin 31.79 - .54.\cos 21.8 + j.54 \sin 21.8) + 1 = 1.78 + j.926$
$= 2 \angle 27.45°$ **Answer**

$E_2 = V_r + I_2.j_2 = 1 + .54 \ -21.8 \ . \ j2 = 1.72 \angle 35.5°$

Prob. 8.2.5. Synchronous generator operation when connected to a large power system.

Example 1

A generator is supplying a large power system which can be represented by its Thevenin's equivalent circuit, E^{th} in series with $X^{th} = j.2$ pu. The generator terminal voltage is 0.97 pu which may be taken as the reference phasor. It delivers a current of 0.8 – j 0.2 pu, and its synchronous reactance is j1 pu. Find (i) the real (P) and reactive (Q) powers into the system at the generator terminals, and

(ii) the generated voltage (Eg) of the generator

(a) for the conditions described above, and

(b) when the generator is delivering the same power (P) to the system, but the generator to terminal voltage is 1 pu.

(Assume that E^{th} is not affected by the change in terminal voltage from 0.97 to 1 pu in case (b)).

Solution. Part (a)

$\mathbf{I} = 0.8 - j.2$ pu

$\text{Re}\,\mathbf{I} = 0.8$

$\text{Im}\,\mathbf{I} = 0.2$

$\mathbf{V} = 0.97$ pu

$P + jQ = \mathbf{VI}$ conjugate $= 0.97(0.8 + j_2)$

$P = 0.776$ pu **Answer**

$Q = V.\text{Im}\,\mathbf{I} = 0.194$ pu **Answer**

$X_s = 1$ pu

$\mathbf{E_g} = \mathbf{V} + \mathbf{I}.jX_s$

$\text{Re}\,\mathbf{E_g} = V + \text{Im}\,\mathbf{I}.X_s = 1.17$ pu

$\text{Im}\,\mathbf{E_g} = \text{Re}\,I.X_s = 0.8$ pu

$E_g = 1.417$ pu **Answer**

Part (b)

$X^{th} = 0.2$ pu

To find P and Q when V = 1 pu we must find the phase angle of V as follows:

$E^{th} = V - j\,X^{th}\,(\text{Re}\,\mathbf{I} - j\,\text{Im}\,\mathbf{I})$

$\text{Re}\,\mathbf{E^{th}} = V - X^{th}.\text{Im}I = 0.93$ pu

$\text{Im}\,\mathbf{E^{th}} = X^{th}\,\text{Re}\,\mathbf{I} = -0.16$ pu

$E^{th} = 0.943$ pu

Angle of $E^{th} = -9.766°$

$P = V\,E^{th}\,\text{Sin}\,\delta.X^{th}$

$$\delta = \sin^{-1}\left(\frac{P.X^{th}}{V.E^{th}}\right) = 9.766°$$

V magnitude $= 1$ pu

angle $= 0$

$\mathbf{I} = V - \mathbf{E^{th}}/jX^{th}$

$$\text{Re}\,\mathbf{I} = \left(\frac{-\text{Im}.E^{th}}{X^{th}}\right) = 0.8 \text{ pu}$$

$$\text{Im}\,\mathbf{I} = - -\frac{V - \text{Re}\,E^{th}}{X^{th}} - 0.35 \text{ pu}$$

$$E_g = \mathbf{V + jXs\ (ReI - jIm\,I)}$$
$$\text{Re}\,\mathbf{E_g} = V - X_s.\text{Im}\,\mathbf{I} = 1.35 \text{ pu}$$
$$\text{Im}\,\mathbf{E_g} = X_s\,\text{Re}\,\mathbf{I} = 0.8 \text{ pu}$$
$$E_g = 1.569$$
$$P + jQ = \mathbf{VI} \text{ conjugate} = 1 \times (0.8 + j0.35)$$
$$P = V.\text{Re}\,I = 0.8 \text{ pu}$$ **Answer**

$$Q = V.\text{Im}\,\mathbf{I} = 0.35 \text{ pu}$$ **Answer**

Specifying a higher terminal voltage at a system bus to which generator is connected results in a larger reactive power supplied to the system by the generator and requires larger E_g obtained by increasing the D.C. field excitation.

	old	**new**
Q	= 0.19	4 0.35 pu
E_g	= 1.42	1.57 pu

Answer

Prob.8.2.6 Power flow on interconnected bus bars

Perform a power flow on the interconnected bus bars shown in Fig.8.2.6

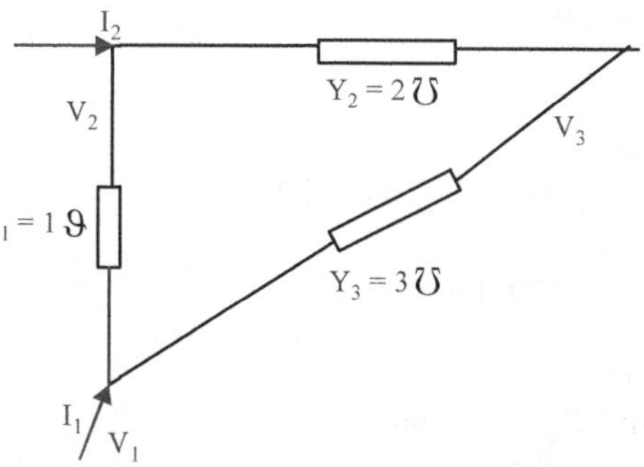

Fig.8.2.6. *Interconnected bus bars*

Solution. The nodal equations are;

I_1, Generator node = 1000 A
I_2 , Load node = – 2000 A
V_3, Slack node = 100 kV, phase value

Solution. The nodal equations are;

$$\begin{pmatrix} I_1 \\ I_2 \\ I_3 \end{pmatrix} \begin{pmatrix} 4,-1,-3 \\ -1,\ 3,-2 \\ -3,\ -2,\ 5 \end{pmatrix} \begin{pmatrix} V_1 \\ V_2 \\ V_3 \end{pmatrix} \qquad \begin{matrix} \ldots(1) \\ \ldots(2) \\ \ldots(3) \end{matrix}$$

Rewrite equation (1) as

$$V_1 = \frac{I_1}{4} + \frac{V_2}{4} + \frac{3V_3}{4} \ldots (4)$$

Substitute equation (1) in (2) and (3)

$$I_2 = \frac{-I_1}{4} + \frac{11.V_2}{4} - \frac{11.V_3}{4} \qquad \ldots (5)$$

$$I_3 = -3.I_1/4 - 11.V_2/4 + 11.V_3/4 \qquad \ldots (6)$$

Also from eqation 5, $V_2 = \frac{I_1}{11} + \frac{4I_2}{11} + V_s$

Substitute for V_2 in the other equations :

$$V_1 = \left(\frac{1}{4}+\frac{1}{44}\right) I_1 + \frac{J_2}{11} + V_3 = \frac{3I_1}{11} + \frac{I_2}{11} + V_3$$

$$V_2 = \frac{I_1}{11} + \frac{4I_2}{11} + V_s$$

$$I_3 = -I_1 - I_2$$

Dealing in kA and kV:

V_1 = 99.909 kV **Answer**

V_2 = 99.364 kV **Answer**

I_3 = generator node = 1 kA **Answer**

Prob. 8.2.7 Synchronous generator operation when connected to a large power system.

Example 2

The generator of Fig.8.2.7(a) operates at rated terminal conditions with a power factor of 0.8 lagging.

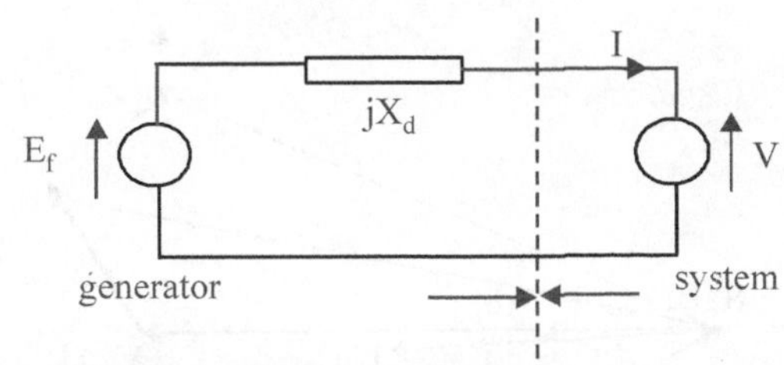

Fig.8.2.7. *(a) Generator connected to a system*

X_d = 0.7 pu on the generator rating

(a) Find P,Q, excitation voltage E_f, and δ. Draw the phasor diagram

(b) Steam valve is opened further so that P increases by 20 %. Re-evaluate P, Q, E_f and δ. Draw the new phasor diagram

(c) The system is restored to the conditions in (a). Exciter is adjusted to raise E_f by 20 %. Re-evaluate P, Q, E_f and δ. Draw the new phasor diagram

Solution. Subscript '0' denotes initial values .Subscript '1' denotes adjusted values:

(a)

V, reference = 1 pu

p.f. = 0.8 lag

$\Phi_0 = \cos^{-1}$ (pf) = 36.869°

I_0 rated current = 1pu

Re $I_0 = I_0 \cos(\Phi_0) = 0.8$ pu

Im $I_0 = I_0 \sin(\Phi_0) = 0.599$ pu

$X_d = 0.7$ pu

$\mathbf{E_{f0} = V + jX_d I_0}$

Re$\mathbf{E_{f0}}$ $= V + X_d I_0 \sin(\phi_0) = 1.419$ pu

Im $\mathbf{E_{f0}}$ $= X_d I_0 \cos(\phi_0) = 0.5$ pu

E_{f0} magnitude = 1.526 pu **Answer**

E_{f0} angle = δ_0 =21.536°. **Answer**

$$P_0 = \frac{1}{X_d}.E_{fo}V \sin\delta_0 = 0.8 \text{ pu}$$ **Answer**

$$Q_0 = \frac{(E_{po}.V\cos\delta_o - V_2)}{X_d} = 0.599 \text{pu}$$ **Answer**

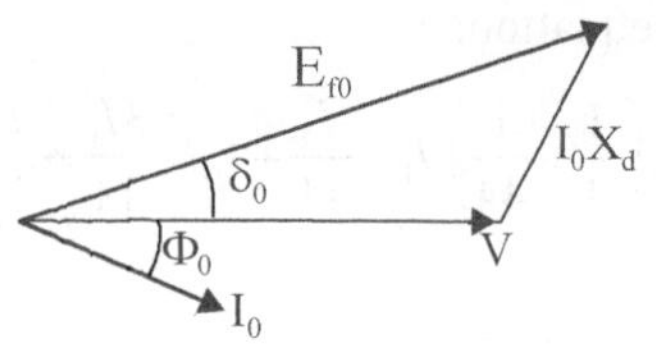

Fig.8.2.7. (*b*) *Phasor diagram for Part*(*a*)

(b)

$P_1 = 1.2\,.P_0 = 0.96$ pu

$E_{f1} = E_{f0}$, no change = 1.53 pu

$$\delta_1 = \sin^{-1}\left(\frac{P_1.X_d}{E_{f1}.V}\right) = 26.067°$$

$$Q_1 = \frac{E_{f1}.V\cos\delta_1 - V^2}{X_d} = 0.535 \text{ pu}$$

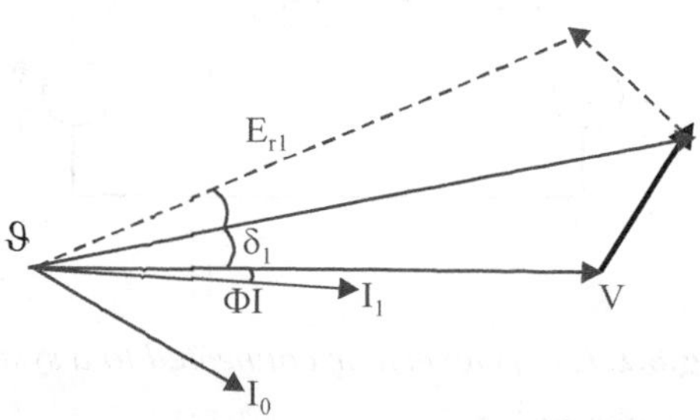

Fig.8.2.7. (*c*) *Phasor diagram for Part* (*b*)

(c)

P_1 = ' P_0' no change = 0.8 pu

$E_{f1} = 1.2\ E_{f0} = 1.831$

$$\delta_1 = \sin^{-1}\left(\frac{P_1X_d}{E_{f1}.V}\right) = 17.812°$$

$$Q_1 = \frac{E_{f1}.V\cos(\delta_1) - V^2}{X_d} = 1.62 \text{ pu}$$

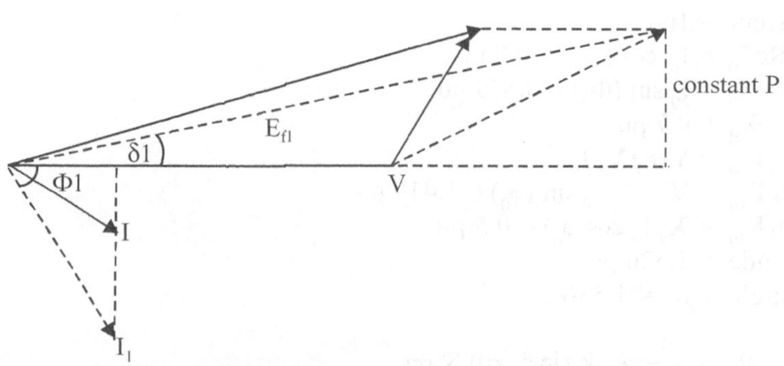

Fig. 8.2.7. (*d*) *Phasor diagram for Part* (*c*).

Prob.8.2.8 Voltage at the bus bar of a lighting load when the induction motors are switched ON

The 132 kV system shown below is represented by a voltage source in series with a 4 % reactance on a 50 MVA base. Fig.8.2.8 shows the equivalent single-phase circuit on 50 MVA base

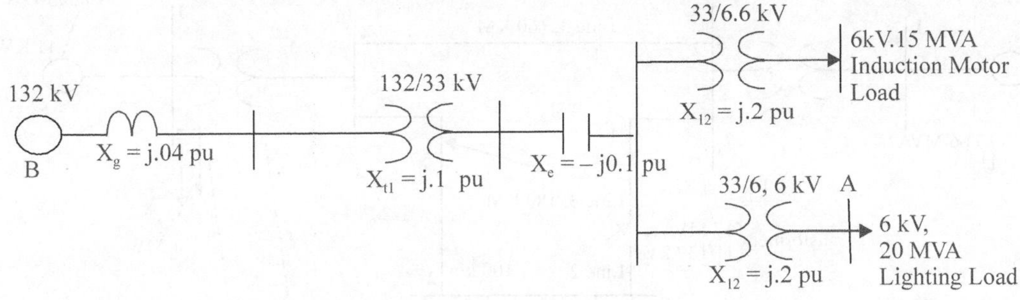

Fig.8.2.8 .*The 132 kV system*

Determine the voltage in pu at bus bar A of the lighting load before and after the induction motor is switched on. The induction motor takes a starting current of 5 times the full-load current at zero power factor lagging.

Solution. I.M. load = 15 MVA

Lighting Load = 20 MVA

Bus bar A voltage = 6 kV

Full-load current of I.M. = 1.443 kA $\left(=\frac{15}{\sqrt{3}\times 6}\right)$

Starting current = 7.216 kA

X_g = 0.04 pu

X_{t1} = 0.1 pu

X_{t2} = 0.2 pu

X_c = – 0.1 pu

Reactance B to A before starting the motor load, X = 0.24 pu

I = Full load current (lighting load) = $\frac{20}{6\sqrt{3}}$ = 1.926 kA

Base MVA = 50

Base current in 6.6 kV circuit = $\frac{50,000}{6.6\sqrt{3}}$ = 4379.051 A

Ipu = 0.44pu @ 0.8 p.f lag power factor (lighting load)

$\phi = \cos^{-1}(0.8) = 35.869^\circ$

Re Ipu = 0.352

Im Ipu = 0.264

V_a at bus bar A , before starting the motors =1– jX(Re Ipu–jIm Ipu)

Re V_a =1– X.I_m Ipu = 0.936 pu

ImV_a = 0.084 pu –

I_1pu = starting current of Induction Motor = – 1.648 pu at 0 pf lagging i.e. – j 1.648

V_a'after starting of motors = 1– j(X_g+ X_{t1}+ X_c).(Ipu(cos Φ – jsin Φ) + I_1pu)– j(Ipu(cos Φ – jsin Φ))X_{t2}

Re V_a' = 1– (X_g+ X_{t1}+ X_c).(Ipu.sin Φ – I_1pu) – Isin Φ.X_{t2} = 0.870

ImV_a' = – Icos Φ (X_{gl} + X_{t1}+ X_{t2}+ X_c) = – 0.084

V_a'mag = 0.874 pu **Answer**

Prob. 8.2.9 Input data & unknowns for a power flow problem

Fig 8.2.9 below shows a 5-bus power system.

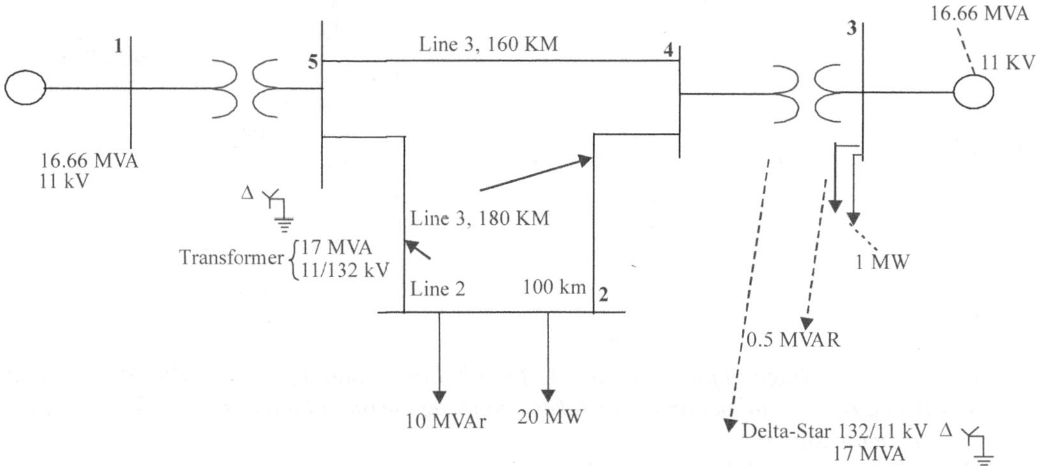

Fig. 8.2.9. *Five-bus power system*

Bus Input Data

BUS	TYPE	Vpu	δdeg	P_G pu	Q_G pu	P_L pu	Q_L pu	$Q_g$maxpu	$Q_g$minpu
1	swing	1	0	–	–	0	0	–	–
2	load	–	–	0	0	0.2	0.1	–	–
3	const.volt	1.05	–	0.1	–	0.01	0.005	0.1	–0.07
4	load	–	–	0	0	0	0	–	–
5	load	–	–	0	0	0	0	–	–

Line Input Data

Bus-to-Bus	Rpu	Gpu	Bpu	Max.MVA pu
2 to 4	0.32	0	0.036	5
2 to 5	0.18	0	0.02	5
4 to 5	0.28	0	0.034	5

Transformer Input Data

Bus-to-Bus	*Rpu*	*Xpu*	*Max.MVA pu*	*Max.Tap Setting pu*
1 to 5	*0.09*	*0.45*	*1.2*	–
3 to 4	*0.09*	*0.4*	*1.2*	–

Note that loads at buses are inductive since $Q_2 = -0.1$ and $Q_3 = -0.005$

For each bus, determine which of the variables are input data and which are unknowns.

Also compute the elements of the second row of Y– bus.

Solution.

Bus	Input Data	Unknowns
1	$V_1 = 1, \delta_1 = 0$	P_1, Q_1
2	$P_2 = -0.2, Q_2 = -0.1$	V_2, δ_2
3	$V_3 = 1.05, P_3 = 0.09$	Q_3, δ_3
4	$P_4 = 0, Q_4 = 0$	V_4, δ_4
5	$P_5 = 0$, Q5= 0	V_5, δ_5

$$\text{ReY}_{24} = \frac{-R_{24}}{R_{24}^{\ 2}+X_{24}^{\ 2}} = -0.136 \text{ pu}$$

$$\text{ImY}_{24} = \frac{X_{24}}{R_{24}^{\ 2}+X_{24}^{\ 2}} = 0.637 \text{ pu}$$

Similarly,

$\text{Re Y}_{25} = -0.249$ pu
$\text{ImY}_{25} = 1.150$ pu
Re Y_{45} = not required
ImY_{45} = not required
$\text{ReY}_{22} = -\text{ReY}_{24} - \text{ReY}_{25} = 0.385$ pu
$\text{ImY}_{22} = -\text{ImY}_{24} - \text{ImY}_{25} + 0.5.B_{24} + 0.5.B_{25} = -1.760$ pu
where half of the shunt admittance of each line is connected and included in Y_{22}.

Prob.8.2.10 Power flow using Gauss-Seidel method

For the power system of Prob. 8.2.9, use Gauss -Seidel method to calculate $V_2(1)$, the voltage at bus 2 after the first iteration. Use zero initial phase angles and 1 pu initial voltage at buses(except at bus 3, where $V_3 = 1.05$) to start the iterations.

Solution: Bus 2 is a load bus.

Ik =(Pk– jQk)/Vk*

$$Vk(i+1) = (1/Ykk)[(Pk-jQk)/Vk^*(i) - \sum_{n=1}^{n=k-1} YknVn(I+1) -?\sum_{n=k+1}^{n=N} YknVn(I)]$$

$V_2(1) = (1/Y_{22})[(P_2 - jQ_2)/(V_2\text{conju}(0)) - [Y_{21}V1(1) + Y_{23}V_3(0) + Y_{24}V_4(0) + Y_{25}V_5(0)]$
$= (1/(0.386 - j\,1.76))[(-0.2 - j(-0.1)/(1) - [0+0 + (-0.136 + j.638)0.1 + (-0.25 + j1.15)1]]$
$= (-0.586 + j1.89)/(0.386 - j1.76)$
$= 1.097\angle 4.85°$

This value is used to recalculate $V_2(1)$

$V_2(1) = (1/(0.386\text{-}j1.76))[(-0.2 + j.1)/(1.097\ \angle -4.85°) - (-0.386 + j1.788)]$ =….. **Answer**

Computations are next performed on buses 3,4 and 5 to complete the first Gauss -Seidel iteration.

Computations are next performed on buses 3,4 and 5 to complete the first Gauss -Seidel iteration.

Prob.8.2.11 Power flow using Newton-Raphson method

For the system shown in Fig. 8.2.11,

$$Y\,bus = \begin{pmatrix} 24.23 & \angle -75.95° & 12.13\angle -104.04° & 12.13 & \angle -104.04° \\ 12.13 & \angle -104.04° & 24.23\angle -75.95° & 12.13 & \angle -104.04° \\ 12.13 & \angle -104.04° & 12.13\angle -104.04° & 24.35 & \angle -75.95° \end{pmatrix}$$

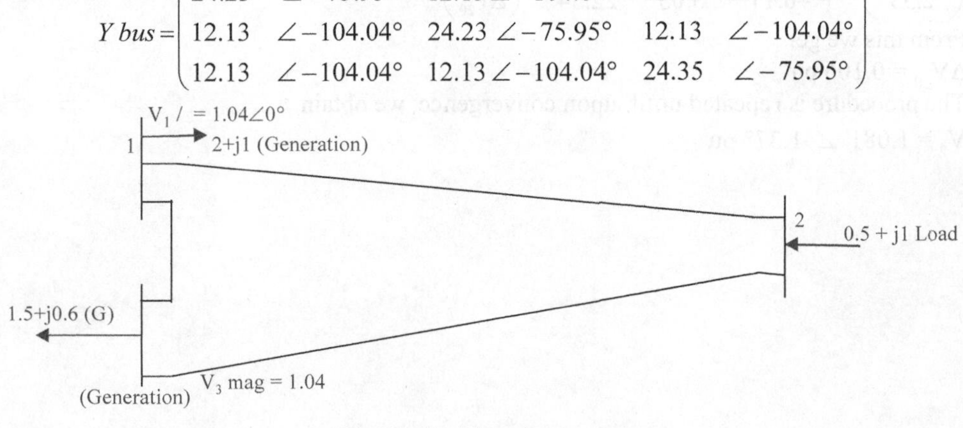

Fig.8.2.11. *Three bus system*

given the pu voltages and powers as shown, determine V_2 by the Newton-Raphson method,\ at the end of the first iteration.

Solution. Let V_{2o}=1 pu, δ_{3o}= 0

$$NPk = \sum_{n=1}^{N} VkVnYkn \cos(\tau kn + \delta n - \delta k) \quad \ldots (1)$$

$$Qk = \sum_{n=1}^{N} VkVnYkn \sin(\tau kn + \delta n - \delta k) \quad \ldots (2)$$

LetV_{2o}= 1

V_{1o}= 1.04

V_{3o}= 1.04

Let

δ_{1o}= 0

δ_{2o}= 0

δ_{3o}= 0

in radians

$$P_{2o} = V_{2o}V_{1o}Y_{21}\cos(\tau_{21} + \delta_{1o} - \delta_{2o}) + V_{2o}^2 Y_{22}\cos\tau_{22} + V_{2o}V_{3o}Y_{23}\,\text{Cos}(\tau_{23} + \delta_{3o} - \delta_{2o}) = 0.33 \text{ pu}$$

$$P_{3o} = V_{3o}V_{1o}Y_{31}\cos(\tau_{31} + \delta_{1o} - \delta_{3o}) + V_{3o}.V_{2o}Y_{32}\cos(\tau_{32} + \delta_{2o} - \delta_{30}) + V_{3o}^2 Y_{33}\cos(\tau_{33}) = 0.026 \text{ pu.}$$

$$Q_{2o} = -V_{2o}V_{1o}Y_{21}\sin(\tau_{21} + \delta_{1o} - \delta_{2o}) - V_{2o}^2 Y_{22}\sin\tau_{22} - V_{2o}V_{3o}Y_{23}\sin(\tau_{23} + \delta_{3o} - \square\delta_{2o}) = -1.33 \text{ pu.}$$

Note minus sign for load,Q_{20}

ΔP_{2o}= P_2sceduled – P_2calculated = 0.5– (– 0.33) = 0.83 pu

ΔP_{3o} = – 1.5 – 0.026 = – 1.526 pu

ΔQ_{2o}= 1– (–1.33) = 2.33 pu

For the 3-bus system, (with V_3 known),

$$\begin{pmatrix} \Delta P_{20} \\ \Delta P_{30} \\ \Delta Q_{20} \end{pmatrix} = \begin{pmatrix} dP_2/d\delta_2 & dP_2/d\delta_3 & dP_2/dV_2 \\ dP_3/d\delta_3 & dP_3/d\delta_3 & dP_3/dV_2 \\ dQ_2/d\delta_2 & dQ_2/d\delta_3 & dQ_2/dV_2 \end{pmatrix} \begin{pmatrix} \Delta\delta_{20} \\ \Delta\delta_{30} \\ \Delta V_{20} \end{pmatrix} \quad \ldots (3)$$

$$\begin{pmatrix} 0.83 \\ -1.526 \\ 2.33 \end{pmatrix} = \begin{pmatrix} 24.47 & -12.23 & 5.64 \\ -12.23 & 24.95 & -3.05 \\ -6.11 & 3.05 & 22.34 \end{pmatrix} \begin{pmatrix} \Delta\delta_{20} \\ \Delta\delta_{30} \\ \Delta V_{20} \end{pmatrix}$$

From this we get

ΔV_{20} = 0.106 pu

The procedure is repeated until, upon convergence, we obtain

V_2 = 1.081 $\angle$–1.37° pu

CHAPTER 9

SHORT CIRCUIT CALCULATIONS

9.1 INTRODUCTION

The reader may have a look at the following Internet websites:

- http://www.ece.utexas.edu/~grady/ABC012.html. This site includes a handy phase sequence calculator. Microsoft FORTRAN Power Station 4.0 source and executable are also included.
- http://www.ece.utexas.edu/~grady/PCFLO.html. This site contains **PCFLOW** – a short circuit, load flow and harmonics program.

9.1.1 Symmetrical Components

These were developed to ease the calculations for unbalanced 3–phase systems and as a help to numerical solution using network analyzers. Even with present day digital computation, the symmetrical components help in solution of unbalanced systems, besides explaining many phenomenon such as rotor heating in machines, neutral current, etc.

The basic concept is to convert a set of three phasors into another set of three phasors with certain desirable properties. The symmetrical components (introduced by Fortescue) is only one such set, the other set is the Kimbark/Clarke components.

The unique property of symmetrical components is that they retain the concept of 3–phase system associated with each component phasor. Thus the positive sequence retains the concept of 3 balanced phasors having the same phase sequence as the original phasors whereas the negative sequence component retains the concept of 3 balanced phasors but rotating in opposite direction of rotation. The zero sequence component is balanced set of 3– coincident phasors but rotating in the same direction as the original unbalanced phasors.

The basic definition for a set of unbalanced three–phase system in terms of the sequence components is

$I_a = I_{0a} + I_{1a} + I_{2a}$

$I_b = I_{0b} + I_{1b} + I_{2b} = I_{0a} + a^2 I_{1a} + a I_{2a}$

$I_c = I_{0c} + I_{1c} + I_{2c} = I_{0a} + a I_{1a} + a^2 I_{2a}$

These equations may be written in matrix form as

$I^{a,b,c} = [T_s] I^{0,1,2}$

where Ts is called the symmetrical component transformation matrix.

$I^{0,1,2} = [Ts]^{-1} I^{a,b,c}$

Similar transformation may be applied for unbalanced voltages also.

Illustrate these transformations by phasor additions.

9.1.2 Power Invariance

If the above transformation T_s is used simultaneously on the voltage and current values of three phase network elements, then

$S_{pq}^{abc} = P_{pq} + j\,Q_{pq} = [(I_{pq}^{abc})^*]^t e_{pq}^{abc}$

And

$S_{pq}^{012} = [(I_{pq}^{012})^*]^t e_{pq}^{012}$

Note that Spq^{abc} is not equal to Spq^{012}.

$S_{pq}^{abc} = 3S_{pq}^{012}$ since S_{pq}^{012} refers to only phase "a" power, and similar amount of power is in phases "b" and "c" also. Thus all three phases of symmetrical components must be used. Often for computer work, the symmetrical transformation given by

$$T_{si} = \frac{1}{\sqrt{3}} T_s$$

is use where $T_{si}^{*t}.T_{si}$ = unity matrix. This is a property of orthogonal matrices. Further since $T_{si}^{*} = T_{si}^{-1}$ we can show that $S_{pq}^{abc} = S_{pq}^{012.}$ However, in earlier works, and even now, the originally defined non–power invariant transformation is being used.

9.1.3 Impedance Transformation

To solve the network in terms of the sequence components, sequence components of impedance are required. These are obtained from their corresponding three phase values.

$E^{abc} = Z^{abc} I^{abc}$

Or,

$T_s E^{012} = Z^{abc} T_s I^{012}$

Or,

$E^{012} = T_{si}^{-1} Z^{abc} Ts\, I^{012}$

$= Z^{012} I^{012}$

Thus $Z^{012} = T_s^{-1} Z^{abc} T_s$

Or,

$Z^{abc} = T_s Z^{012} T_s^{-1}$

Forms of Z^{012} for balanced stationary and rotating elements should be known. Decoupling of sequence components and its limitations are to be stressed.

Sequence generated voltages in a balanced network are :

$E^1 = E_a, E^2 = 0, E^0 = 0$

9.1.4 Sequence Networks

These networks have the advantage since for balanced network there is no mutual coupling between sequence component elements unlike what happens in balanced 3-phase components. Thus, balanced 3-phase network can be assembled component for component in three separate sequence networks. Sequence networks of generator, transmission line, transformer, and loads should be known. Zero sequence networks for different transformer connections should also be known. The ideal earth and the neutral point should be distinguished. Relative magnitudes of sequence impedances of generator and transmission lines should be known.

1. Find Z^{012} for rotating and non-rotating elements in terms of corresponding phase quantities

2. Study the measurements of zero sequence impedances of transformer for different connections

3. Find phase shifts in sequence components in ΔY transformation

4. If the positive, negative and zero sequence impedances of a transmission line are 0.3, 0.3, and 0.5 respectively, find the self and mutual impedances between the phases.

9.1.5 Fault Analysis

An essential part of the design of a power system is the calculation of currents which flow in the components and the resulting voltages, when the faults of various types occur. Common faults on a transmission system are;

1. LG
2. LL
3. LLG

4. LLL or LLLG
5. Open conductor
6. Simultaneous faults that may be any combination of the above five.

Faults may also occur in switchgear, transformers and machines but their nature may be different than those in transmission lines.

9.1.6 Causes of System Faults

Faults are mostly due to lightning and switching. Most of them are temporary. LG fault is most common and LLL fault is least common. Relays provided on the system detect these faults and produce a trip signal for the circuit breaker to isolate the faulty portion form the system. Thus to determine the circuit breaker and relay operating times, fault currents & voltages have to be calculated and for many other applications.

Except for the three phase faults, all other types of faults cause unbalanced operation, and fault currents & voltages under such conditions are required to be obtained using symmetrical components, or phase components (the latter analysis is much more difficult even with digital computers.

The calculation of 3-phase balanced faults is relatively simple but forms the basis of determining the circuit breaker ratings.

9.1.7 Symmetrical Faults

When a sudden short circuit occurs on the electric supply system, the currents & voltages are of transient nature before they settle down to steady state values.

The fault current at any time consists of :

1. DC transient , also called the DC offset, arising out of the terms of the type $Ae^{-Rt/L}$
2. AC transients consisting of the terms of the type $Ae^{-Rt/L}$ Sin $(\omega t + \phi)$ where ω= supply frequency
3. Steady state value of the type A sin $(\omega t + \phi)$.

For a single machine system, the maximum value of the DC transient could be equal to the peak value of the total AC component. The symmetrical current in one phase will have no DC offset. (Other phases will have DC offset). The cause of sub-transient/transient current could be explained on the basis of the theorem of constant flux linkages associated with the field winding. The damper winding affects the initial value of the fault current. The initial value decays faster because of the very low time constant of this circuit. Later, the closed field circuit effects dominate.

9.1.8 Definition of Reactances

$$X_d'' = \frac{\sqrt{2}E}{O_a}$$

$$X_d' = \frac{\sqrt{2}E}{O_b}$$

$$X_d = \frac{\sqrt{2}E}{O_c}$$

What are the typical values of these reactances for turbo and hydro generators ?

For the interruption current rating of a circuit breaker (time involved being 2-5 cycles), the sub-transient current is important. The transient current is important for transient stability studies lasting from 1-2 seconds to 10 seconds.

Short circuit volt-amperes = $\sqrt{3}$. (short-circuit current). [Nominal Voltage (line value)]

9.1.9 System Representation During Short Circuit

Simplifying assumptions :

- Each machine is represented by a constant voltage behind the machine reactance, sub-transient or transient depending on the situation.
- Shunt connections i.e., loads, line charging, etc., are neglected.
- All transformers are assumed to be at nominal tap settings.
- The above three assumptions do not imply that the system is unloaded before the fault. However, it is assumed that the network is unloaded and that all the generator voltages are at 1 pu. This assumption is justified because fault currents in the network are much larger than the load currents.

If assumptions 2-4 are not taken into account, pre-fault currents are added to the fault currents obtained by using the Thevenin's equivalent of the network.

Thevenin's equivalent

The Thevenin's equivalent of the network at the point of the fault location (either in phase or sequence components) is obtained for calculating the fault currents. If the network is balanced before the fault, then at the fault point, Thevenin's sequence voltages become

$E_{a1} = E_a$

$E_{a2} = 0$

$E_{a0} = 0.$

If the network is not balanced, the above voltages may have finite values and the sequence component approach may not be useful. The performance equation for the balanced network in terms of sequence components is:

$[V_s] = [E_{ph}] - [Z_s]\,[I_s],$

where, $[V_s] = [V_{0a}, V_{1a}, V_{2a}]$

$[E_{ph}] = [0, E_{1a}, 0]$

$[Z_s]$ = diagonal matrix of $[Z_0, Z_1, Z_2]$

$[I_s] = [I_{0a}, I_{1a}, I_{2a}]$

The network at the fault point F may be drawn, where Z_1, Z_2 and Z_0 are the lumped values of the Thevenin's impedances between the fault point and the neutral.

Depending upon the type of fault, the sequence components of currents and voltages are constrained leading to particular connections of sequence network. After the fault currents coming out of the network at the fault point are determined, the current & voltage distributions inside the network are found out. If originally, there are load currents, these are restricted to the positive sequence network, and they are superposed on the fault currents for accurate results, but this is often not done.

9.1.10 Three-phase fault

Here,
$$I_{a1} = \frac{E_a}{(Z_1 + Z_F)}$$
$$I_{a2} = 0$$
$$I_{a0} = 0$$

Students must be thoroughly familiar with procedures for calculating 3-phase faults. For a given fault, such calculations include the determination of the voltage at any point in the network, the currents in any branch and the value of reactance necessary to limit the fault current to any desired value.

Such problems may be solved by (*i*) reducing the network to a single equivalent impedance, or (*ii*) for more extensive networks, by treating the problem as a special case of a load-flow problem and using matrix and computer techniques.

Network reduction usually involves star/delta or delta/star transformation.

9.1.11 Single-line to ground (SLG) fault

Fault is assumed in phase "a". If it is in any other phase, symmetrical shifting will be required. The connections at the fault point can be drawn for various types of faults.

9.1.12 LL and DLG faults

You may look up the network connections for L-L and DLG faults and answer the following questions.

1. How do you distinguish switching transient currents from sub-transient currents?

2. Sketch the sub-transient/transient current wave-shapes in phase *b* and *c* respectively for a synchronous generator when the current in phase a is symmetrical.

3. Under what conditions can you expect a negative or zero sequence Thevenin's voltage at the fault point ?

(*a*) Determine and draw the sequence network connection diagrams for

- Single conductor opening at a point
- Two conductors opening at a point.

In the case of SLG and LL faults the expressions for the fault current, $I_F = 3E/ (Z_1 + Z_2 + Z_0)$ and $3E/ (Z_1 + Z_2)$ respectively may be used. Network reduction techniques are used, as with the symmetrical fault, to find the single equivalent values of the impedances of each of the component networks, particular attention being paid to that for Z_0 which depends on the transformer connections and earthing arrangements.

By substituting the boundary conditions relevant to a particular type of fault (e.g. $I_a = 0$ and $V_b = V_c = 0$ for a DLG fault), using the symmetrical components analysis, relations between the component currents and voltages can be found. This leads to the interconnected network method of treatment in which inspection of the component relations enables the sequence networks to be interconnected in such a way as to satisfy the relations (networks in series for a SLG fault and in parallel for LL and DLG faults); the component currents and voltages can then be found by simple circuit calculations and the actual value determined by using the relationships between actual quantities and sequence quantities.

Although calculations are simplest if the reference phase (phase a) is chosen so that it is symmetrically placed with regard to the fault (*i.e.* fault on phase a for a SLG fault and on *b* and c for LL and DLG faults).

OBJECTIVE TYPE QUESTIONS

1. If a positive sequence current passes through a transformer and its phase shift is 30°, the negative sequence current flowing through the transformer will have a phase shift of

(*a*) 30°

(*b*) – 30° (*c*) 120°.

Ans. (*b*)

2. The zero sequence impedances of an ideal star-delta connected transformer (star-grounded)

(a) looking from star side is zero and looking from delta side is infinite.

(b) looking from star side is zero and looking from delta side is also zero.

(c) looking from star side is infinite and looking from delta side is zero.

Ans. (*a*)

3. Symmetrical component method of analysis is more useful when

(a) system has unsymmetrical fault and the network is otherwise balanced.

(b) system has symmetrical fault and the network is otherwise unbalanced.

(c) system has unsymmetrical fault and the network is unbalanced.

Ans. (*a*)

4. For a line to line fault analysis using symmetrical components,
   (*a*) the positive and negative sequence networks at the fault point are connected in series.
   (*b*) the positive and negative sequence networks at the fault point are connected in parallel.
   (*c*) the positive, negative, and zero sequence networks at the fault point are connected in parallel.

   Ans. (*b*)
5. The machine reactances used for computation of short circuit current ratings of a circuit breaker are
   (*a*) synchronous reactance
   (*b*) transient reactance (*c*) sub-transient reactance

   Ans. (*c*)
6. The load currents in short-circuit calculations are neglected because
   (*i*) short-circuit currents are much lager than load currents
   (*ii*) short -circuit currents are greatly out of phase with load currents
   The correct alternative is
   (*a*) both (*i*) and (*ii*) are wrong (*b*) (*i*) is wrong and (*ii*) is correct
   (*c*) both (*i*) and (*ii*) are correct.

   Ans. (*c*)
7. Which one of the following statements is *not* true?
   (*a*) Fault levels in an all a. c system are less than in an a. c system with a D.C. link operating
   (*b*) Large systems may be interconnected with d. c link of small capacity
   (*c*) Limitation on the critical length of underground cables for use in A. C no longer exists if D.C transmission by cables is used.
   (*d*) Corona loss and radio and TV interference with D.C. transmission is less.

   Ans. (a)
8. Advantage of power system interconnection is.
   (*a*) Large size circuit breakers are required because of large short-circuit currents.
   (*b*) Machines of one system remain in step with machines of another system.
   (*c*) Fewer machines are required as reserve for operation at peak loads.

   Ans. (*c*)
9. The internal voltages of a 3-phase synchronous generator correspond to
   (*a*) Positive sequence
   (*b*) Negative sequence
   (*c*) Zero sequence

   Ans. (*a*)

PROBLEMS WITH SOLUTIONS

Prob.9.2.1 Short-circuit level of 4 alternators in parallel

Four identical alternators, each rated at 11 kV, 25 MVA, and each having a sub-transient reactance of 0.16 Ω on its rating, are connected in parallel to a common bus bar. Find the 3-phase fault level at the outgoing feeder from the common bus bar.

Solution: Base kV = 11
Base MVA = 25
X′pu = 0.16
pu fault MVA from one alternator = 1/X′pu
pu fault MVA from one alternator = 6.25 pu(MVA)
Fault MVA = 156.25 MVA
Number of alternators = 4
Total fault MVA = $156.25 \times 4 = 625$ MVA

Prob.9.2.2 Fault current-single-phase fault on an alternator

A 3- phase 33 kV,37.5 MVA alternator is connected to a 33 kV overhead line which develops an earth fault at the remote end. The positive, negative and zero-sequence reactances of the line are 6.3, 6.3, and 12.6 Ω/conductor and those for the alternator are 0.18, 0.12 and 0.1 pu respectively. Calculate the fault current in kA, and the phase voltages at the alternator terminals in kV. Assume the alternator star point is solidly earthed.

Solution. Base MVA = 37.5

Base kV = 33

Base I = 0.656 kA

Base X = 29.04 Ω

Alternator sequence reactance's

$X+pu = 0.18$

$X-pu = 0.12$

$X_o pu = 0.1$

$X+ = 6.3\ \Omega$

$X- = 6.3\ \Omega$

$X_o = 12.6\ \Omega$

$X+pu = 0.216$

$X\text{-}pu = 0.216$

$X_o pu = 0.433$

Total reactance of three sequence networks in series = 1.267 pu

$E = I+ = I- = Io = E/\text{Total reactance} = 0.788$ pu

Fault current = $3I+ = 2.366$ pu = 1.554 kA

Voltages w. r. t earth at alternator terminals are calculated using

$V+ \;= E - I_+Z_1 = 0.858$ pu

$V- \;= -I_-Z_2 = 0.094$ pu

$V_o \;= -I_oZ_o = -0.078$ pu

$V_a \;= V_+ + V_- + V_o = 0.684$ pu

$V_b \;= a^2V_+ + aV_- + V_o$

$\text{Re}V_b \;= -0.5\,V_+ - 0.5V + V_0 = -0.460$

$\text{Im}V_b \;= 0.866\,(V_+ - V_-) = 0.825$

$V_b(\text{mag}) = 0.944$ pu

$= 18.002$ kV

V_b angle = – 60.827°

$\text{Re}V_c \;= -0.5\,V+ - 0.5\,V\_ + V0 = -0.460$

$\text{Im}V_c \;= 0.866\,(-V_+ + V_-) = -0.825$

$V_c(\text{mag}) = 0.944$ pu

$= 18.002$ KV

V_c angle = 60.827°

Prob.9.2.3 Short-circuit level for a fault on an 11 kV feeder of a power system

Calculate the fault level for a 3-phase symmetrical short-circuit on an 11 kV feeder in the power system of Fig.9.2.3 (a).

Generator reactance = 0.16 pu

Generator MVA = 25

Transformer reactance = 0.125 pu

Transformer MVA = 50
Reactor reactance = 0.053 pu
Reactor MVA = 25
Grid in feed = 1500 MVA
Base MVA = 100
Grid in feed on base MVA = 15 pu
E Grid voltage = 1 pu
EI Grid volt-amp = 15 pu

$$\text{Grid source reactance on (pu)} = \frac{E}{I} = \frac{E^2}{E.I} = 0.066\text{pu}$$

Generator reactance = 0.64 pu
Transformer reactanceon common base = 0.25 pu
Reactor reactance on common base = 0.212 pu

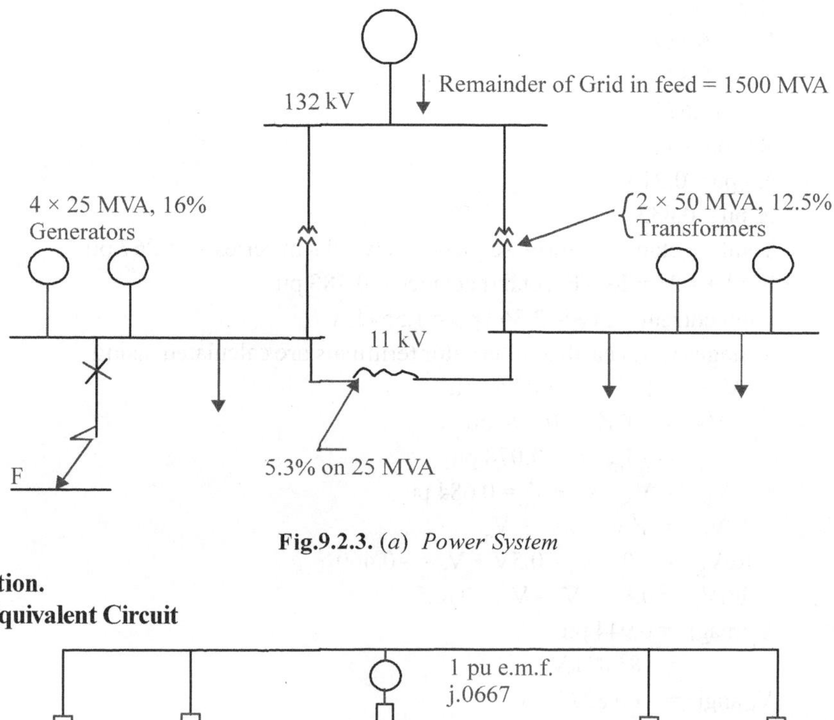

Fig.9.2.3. (*a*) *Power System*

Solution.

pu Equivalent Circuit

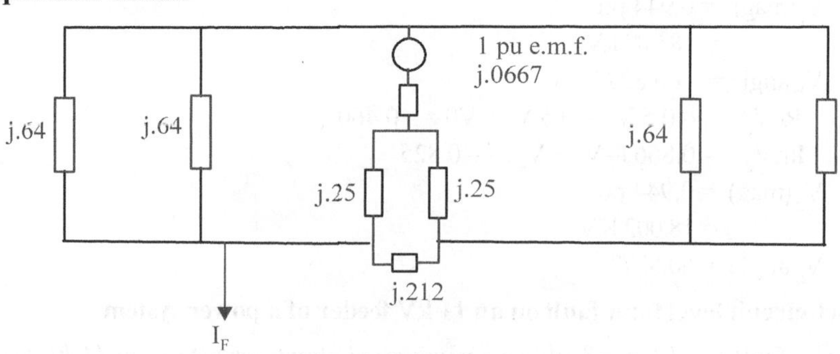

Fig.9.2.3. (*b*) *pu equivalent circuit*

Reduce the circuit to Fig.9.2.3 (c) by combining two parallel branches and converting delta into star. Further reductions give Fig.9.2.3 (d) and Fig . 9.2.3 (e) respectively.

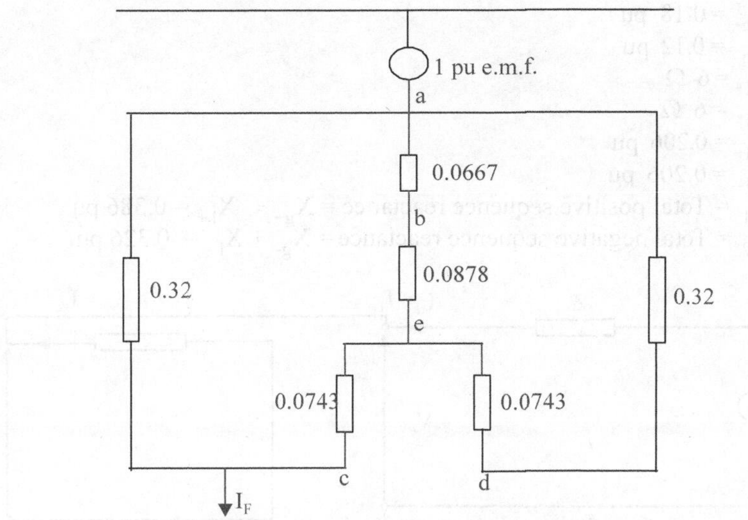

Fig.9.2.3. (*c*) *Reduced equivalent circuit*

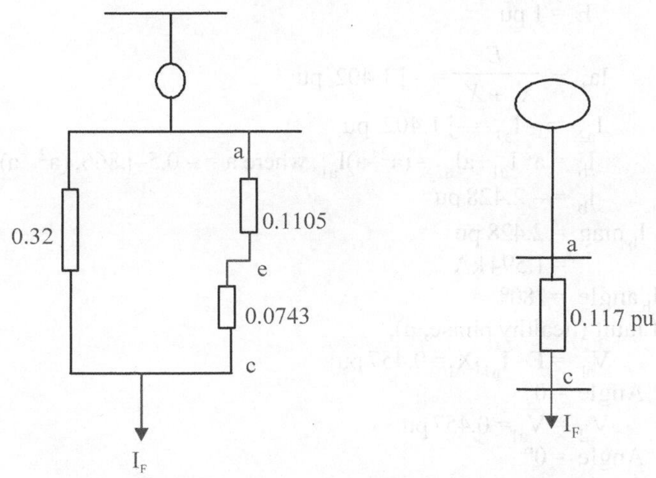

Fig. 9.2.3 (d). *Reduced Equivalent circuit from Fig. 9.2.3.(c)*

$$\text{Fault MVA} = \frac{100}{0.117}$$

$= 854.7009$ MVA **Answer**

The circuit breaker will thus be rated at the next standard size 11 kV, 1000 MVA

Prob.9.2.4 Fault current and the voltage at the healthy phase for a line-line fault at the remote end of a line connected to an alternator

A 3-phase 33 kV 37.5 MVA alternator is connected to a 33 kV overhead line which develops a short-circuit between phases b and c at the remote end. The positive and negative sequence reactances are 18 and 12 % (on rating) while those for the line are 6.3 Ω each. Calculate the fault current , and the voltage of the healthy phase to earth at the fault point.

Solution. Base MVA = 37.5

Base kV = 33

Base X = 29.04 Ω

$X_{g+} = 0.18$ pu
$X_{g-} = 0.12$ pu
$X_{l+} = 6\ \Omega$
$X_{l-} = 6\ \Omega$
$X_{l+} = 0.206$ pu
$X_{l-} = 0.206$ pu
X_1 = Total positive sequence reactance = $X_{g+} + X_{l+} = 0.386$ pu
X_2 = Total negative sequence reactance = $X_{g-} + X_{l-} = 0.326$ pu

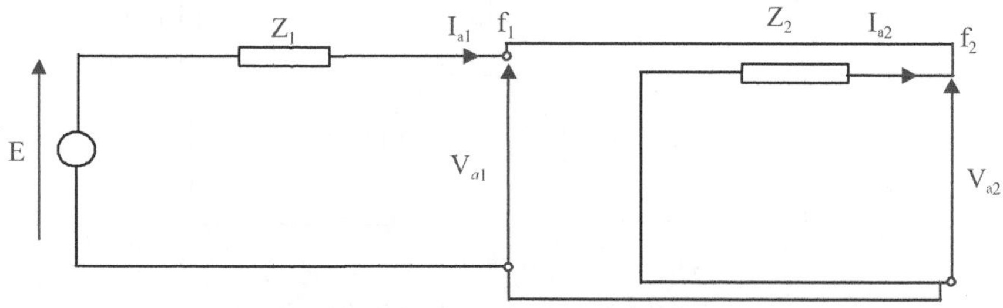

Fig. 9.2. 4. *Sequence network connections for a line to line fault*

$$E = 1 \text{ pu}$$

$$Ia_1 = \frac{E}{X_1 + X_2} = -j\ 1.402 \text{ pu}$$

$$I_{a2} = -I_{a1} = -j\ 1.402 \text{ pu}$$

$$I_b = a^2 I_{a1} + aI_{a2} = (a^2 - a)I_{a1}, \text{ where } a = -0.5 - j.866,\ (a^2 - a) = -j1.732$$

$$I_b = -2.428 \text{ pu}$$

$$I_b\text{mag} = 2.428 \text{ pu}$$

$$= 1.594 \text{ kA}$$ **Answer**

$$I_b\text{ angle} = 180°$$

At the point of fault (healthy phase, a),

$$V_{al} = \mathbf{E} - \mathbf{I_{a1}} jX_1 = 0.457 \text{ pu}$$

$$\text{Angle} = 0°$$

$$V_{a2} = V_{a1} = 0.457 \text{ pu}$$

$$\text{Angle} = 0°$$

$$V_{æ} = V_{a1} + V_{a2} = 0.915 \text{ pu}$$

$$= 17.449 \text{ kV}$$ **Answer**

$$\text{Angle} = 0°$$

$$\text{Fault MVA} = 1.594 \times 33 = 52.63$$ **Answer**

52.630 MVA

Prob.9.2.5 Generator supplying power to a synchronous motor via a transmission line- To find fault currents in the machines for a 3-phase fault at the alternator terminals

A 22 kV 100 MVA alternator with 0.25 pu transient reactance is supplying a load through a transmission line of reactance 0.05 pu at 100 MVA. The load at a particular time is equivalent to a 50 MVA synchronous motor with 0.2 pu transient reactance which is taking 40 MW at .8 pf leading with a terminal voltage of 21.9 kV. If a 3-phase short-circuit occurs at the alternator terminals, calculate the current in each of the two machines and in the fault during the transient period.

Solution. MVA_{b}, base value = 100
kV_{b}, base value = 22
I_{b}, base value = 2624.321 A
P_{m}, motor load = 40 MW

V_{m}, motor voltage = 21.9 kV
= 0.995 pu
p.f.m. motor p.f. = 0.8 lead
I_{m}, motor current = 1318.152 A
= 0.502 pu
Motor MVA = 50
X_{m}, on equipment base = 0.2 pu
X_{m}', on 100 MVA base = 0.4 pu
X_{g}', on 100 MVA = 0.25 pu
X_{l}, line reactance = 0.05 pu

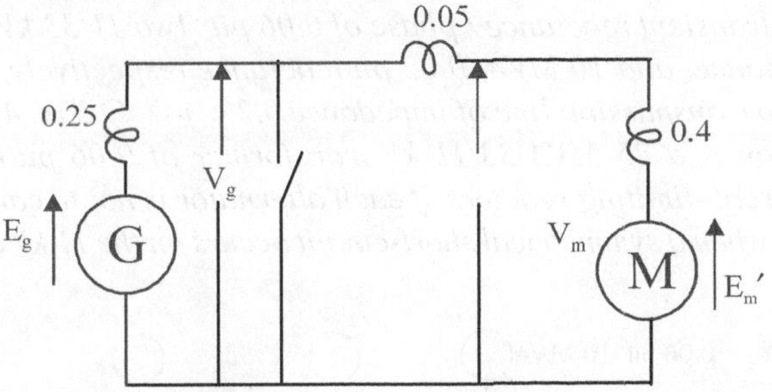

Fig.9.2.5. *Power system*

Vm is the reference phasor

$\text{Re}\,\mathbf{I_m} = I_m \times \text{p.f.m.} = 0.401$ pu

$\text{Im}\,\mathbf{I_m} = I_m \times \sin(\cos^{-1}(\text{p.f.m.}) = 0.301$ pu

$\mathbf{V_g} = V_m + jX_l(\text{ReIm} + j\text{Im Im})$

$\text{Re}\,\mathbf{V_g} = V_m - Xl^*\text{Im Im} = 0.980$ pu

$\text{Re}\,\mathbf{E_g}' = \mathbf{Re}V_g - \mathbf{X_g}^*\mathbf{ImI}_m = 0.905$ pu

$\text{Im}\,\mathbf{E_g}' = \mathbf{ImV_g} + \mathbf{Xg}' \times \mathbf{Re\,I_m} = 0.120$ pu

$\mathbf{E_m}' = V_m - jX_m'^*(\text{Re}\mathbf{I_m} + j\text{Im}\mathbf{I_m})$

$\text{Re}\,\mathbf{E_m}' = V_m + X_m.\text{Im}\,\mathbf{I_m} = 1.116$ pu

$\text{Im}\,\mathbf{E_m}' = -X_m \times R_e I_m = -0.160$ Pu

Total fault current during the transient period is found by superposition of I_g' and I_m' which flow through the switch.

$I_g' = \dfrac{E_g'}{jX_g'}$

$\text{Re}\,I_g' = 0.482$ pu
1265.426A

$\text{Im}\,I_g' = -3.620$ pu
-9500.5 A

$I_m' = \dfrac{E_m'}{jX_m' + jX_1}$

$\text{Re}\,I_m' = -0.357$ pu
$= -937.352$ A

$\text{Im}\, I_m'$ $= -2.480$ pu

$= -6508.33$ A

Total current in each phase of the short-circuit:

Real part = 328.073 A

Imaginary Part = –16008.8 A

Prob.9.2.6 To find the reactance of the current limiting reactors to limit the fault current

Three star-connected 11 kV alternators are connected each in series with a similar current-limiting reactor to a common bus bar as shown in Fig. 9.2.6(a). The alternators each have a rating of 10 MVA and sub-transient reactance / phase of 0.06 pu. Two 11/33 kV transformers of 15 MVA rating, 0.03 pu reactance, and 10 MVA, 0.02 pu reactance respectively, connected in parallel to this bus bar, supply a transmission line of impedance 0.2 + j .7 Ω /km. At a substation 10 km from the generating station is a 25 MVA 33/11 kV transformer of 0.06 pu reactance. Calculate the reactance of the current -limiting reactors if each alternator is not to carry more than 2.333 times the full-load current, when a symmetrical short-circuit occurs on the 11 kV bus bars in the substation.

Solution.

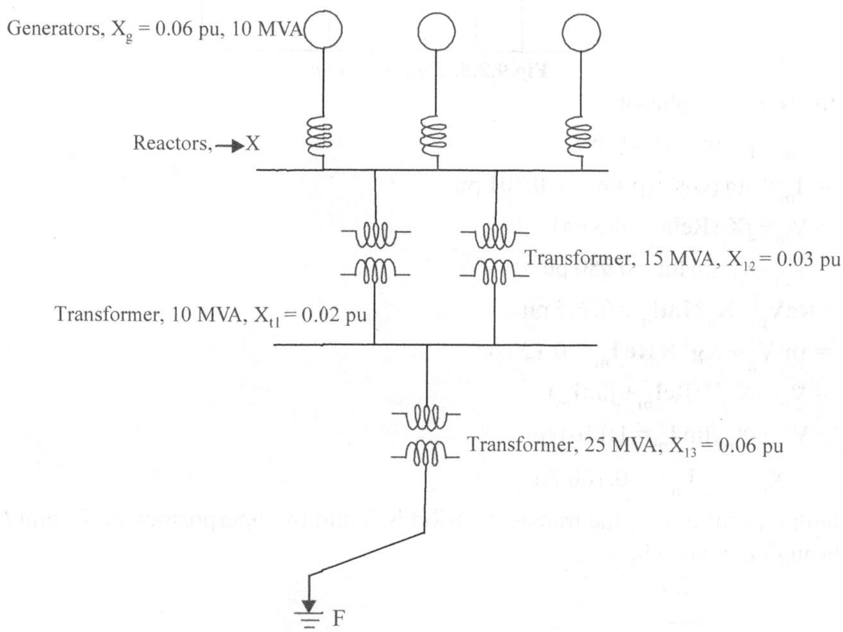

Fig.9.2.6. (*a*) *Power system*

MVA_b	= 25
l	Line length = 10 km
R_l'	line resistance/km = 0.2 Ω/km
X_l'	line reactance /km = 0.7 Ω/km
R_l	$= R_l' \times l = 2\,\Omega$
X_l	$= X_l' \times l = 7\,\Omega$
V_b	in line = 33 kV

$$R_l\text{pu} = \frac{X_1 \times MVA_b}{V_b^2} = 0.045 \text{ pu}$$

$$X_l\text{pu} \quad \frac{X_1 \times MVA_b}{V_b^2} = 0.160 \text{ pu}$$

X_g', on generator base = 0.06 pu

$X_{g'}$ on common base $= \dfrac{0.06\times25\times11\times11}{10\times33\times33} = 0.016$ pu

MVA_{t1}, transformer t1 rating = 10 MVA
MVA_{t2}, transformer t2 rating = 15 MVA
MVA_{t3}, transformer t3rating = 25 MVA
X_{t1}, on common base = 0.05 pu
X_{t2}, on common base = 0.05 pu
X_{t3}, on common base = 0.06 pu

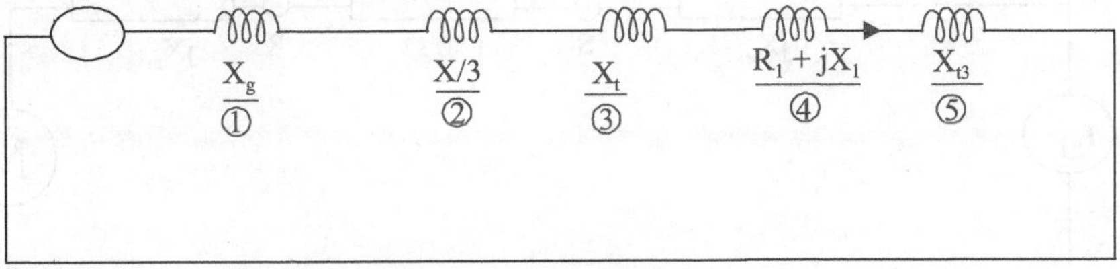

Fig.9.2.6. (*b*) *pu Reactance diagram*

① = generator reactance of 3 generators in parallel = X_g
② = reactance of 3 reactors in parallel = X/3, X is the unknown
③ = reactance of two transformers in parallel = $X_t = \dfrac{X_{t1}.X_{t2}}{X_{t1}+X_{t2}}$
④ = line Impedance = $R_1 + jX_1$
⑤ = transformer reactance = X_{t3}

Total reactance = 0.251+X/30.25
Total impedance = 0.0459 + j(.251+X/3)

Fault current in the pu circuit $= \dfrac{1}{0.0459 + j(0.251 + x/3)}$

I_{mag} = Magnitude of fault current $= \dfrac{1}{\sqrt{(0.459)^2 + \left(0.251 + \dfrac{x}{3}\right)^2}}$

Fault current in each alternator = Imag/3
MVA rating of each alternator = 10 , kV rating = 11
Full load current of each alternator = MVA rating/1.74* kV rating = 524.864 A

I_b, base current $= \dfrac{25\times100}{33\sqrt{3}}$ =437.386 A437.386A
Full load current of each alternator, pu = 1.2
Therefore this value times (7/3) should be equal to Imag/3 .
or,

$$0.459^2 + (0.251 + \frac{X}{3})^2 = \left(\frac{1}{8A}\right)^2$$

or,
X = 0.423 pu
XΩ = 18.446

Answer

Prob.9.2.7 A generator supplying power to a motor: For the single phase fault at the generator terminals, to find the fault current(in the faulted phase) flowing into the fault and the contributions of the generator & motor to the fault.

In Fig. 9.2.7(a), G is an exporting grid area(generator) having a 3-phase fault level of 20000 MVA at the 400 kV bus bars S. The corresponding data for the grid importing M (motor) is 10000 MVA at R. The short 400 kV line has a series reactance of 40 Ω / phase. The load transfer is 150 MW, 0.8 p.f. lagging, 360 kV at R. If a short circuit to earth occurs on one phase at S, calculate the current (in phase A) flowing into the fault and the contributions of the generator & motor to the fault current (in phase A).

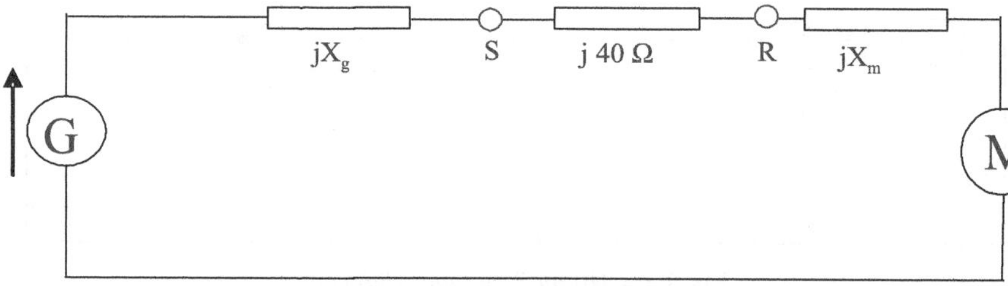

Fig.9.2.7. (a) Power system

For both areas during the sub-transient period, the negative and zero sequence reactances may be assumed to be 100 % and 50 % respectively of the positive sequence values for the grid areas and 100% and 20% for the line. Both G and M are solidly earthed.

Solution. $MVA_{g,}$ fault, MVA of gen.= 20000 MVA

$V,$ bus bar Voltage = 400 kV

$I_{g'}$ Gen. fault current = 28.901 kA

$X_{g'}$ Source reactance = 7.990 Ω / phase

$MVA_{m,}$ for motor = 10000 MVA

$V,$ for motor = 400 kV

$I_{m'}$ for motor = 14.450 kA

$X_{m'}$ for motor = 16 Ω/ph

Take V_R (voltage at R) as reference phasor

V_R L–L = 360 kV

L–N = 207.846 kV/ph

P Load power = 150 MW

load p.f. = 0.8 lag

Im load current magnitude = 0.301 kA

$\text{Re}\,\mathbf{I_m} = 0.240$ kA

$\text{Im}\,\mathbf{I_m} = 0.180$ kA

X_l line reactance = 40 Ω

$\mathbf{V_S} = V_R + \mathbf{ImI_m} * \mathbf{jX_l} = 208.026$ kV/ph

$\text{Re}\,\mathbf{V_S} = 215.071$ kV/ph

$\text{Im}\,\mathbf{V_S} = \mathbf{ReI_m} * X_l = 9.633$ kV/ph

This is the balanced pre-fault voltage at S ....= V_f

The following are generator, line and motor sequence reactances :

$X_{g+} = 7.990$ Ω

$X_{g-} = 7.990$ Ω

$X_{g0} = 3.995$ Ω

$X_{l+} = 40\ \Omega$

$X_{l-} = 40\ \Omega$

$X_{l0} = 8\ \Omega$

$X_{m+} = 16\ \Omega$

$X_{m-} = 16\ \Omega$

$X_{mo} = 8\ \Omega$

The following shows the sequence network for a single –line to ground fault.

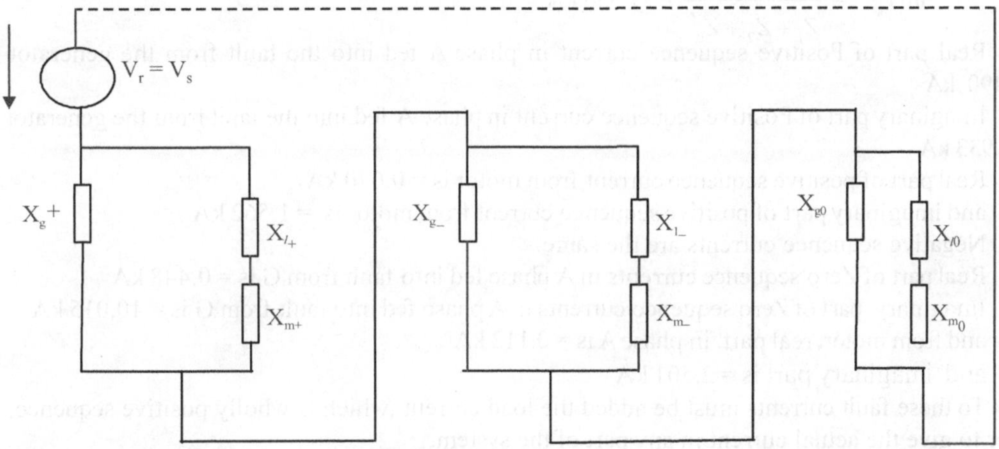

Fig.92.7. (b)Equivalent circuit

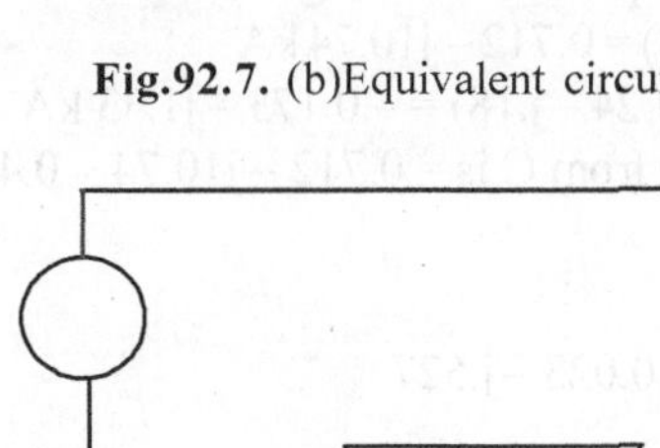

$$Z_1 = \frac{X_{g+}.(X_{1+} + X_{M+})}{X_{g+} + X_{1+} + X_{M+}}$$

$$Z_2 = \frac{X_{g-}(X_{1-} + X_{M-})}{X_{g-} + X_{1-} + X_{M-}}$$

$$Z_0 = \frac{X_{go}(X_{10} + X_{MO})}{X_{go} + X_{10} + X_{MO}}$$

Fig.9.2.7. (c) Reduced equivalent circuit

$Z_1 = 6.992\ \Omega$

$Z_2 = 6.992\ \Omega$

$Z_0 = 3.196\,\Omega$

Fault current flowing from A phase conductors into the fault at S is

$$I_{ae} = \frac{3V_f}{Z_1 + Z_2 + Z_0}$$

$$\text{Re}\, I_{ae} = \frac{3\,\text{Im}\, V_f}{Z_1 + Z_2 + Z_0}$$

$$\text{Im}\, I_{ae} = \frac{3\,\text{Re}\, V_b}{Z_1 + Z_2 + Z_0'} = 37.550 \text{ kA}$$

Real part of Positive sequence current in phase A fed into the fault from the generator is = 0.490 kA

Imaginary part of Positive sequence current in phase A fed into the fault from the generator is = 10.933 kA

Real part of positive sequence current from motor is = 0.070 kA

and imaginary part of positive sequence current from motor is = 1.562 kA

Negative sequence currents are the same.

Real part of Zero sequence currents in A phase fed into fault from G is = 0.448 kA

Imaginary part of Zero sequence currents in A phase fed into fault from G is = 10.015 kA

and from motor, real part, in phase A is = 0.112 kA

and imaginary part is = 2.501 kA

To these fault currents must be added the load current ,which is wholly positive sequence, in order to give the actual current in any part of the system.

The positive sequence current in phase A flowing from G is

$= 0.49067 - j10.953 + (0.24 - j0.18) = 0.712 - j10.74$ kA

and from M is $= 0.07 - j10.56 - (0.24 - j.18) = -0.173 - j1.33$ kA

Actual current in phase A flowing from G is $= 0.712 - j10.74 + 0.472 - j10.56 + 0.517 - j11.57$

$= 1.7 - j32.87$ kA

and from motor is

$= -0.173 - j1.33 + 0.067 - j1.51 + 0.023 - j.527$

$= -0.083 - j3.367$ kA

or $.083 + j\,3.367$ to motor

Sum of these actual currents in phase A is

$= 1.701 - j32.87 - 0.083 - j3.37$

$= 1.618 - j\,36.24$ kA, that is the actual fault current.

Prob.9.2.8 Line-to-line fault at the terminals of a generator-To find fault current

The positive ,negative, and zero sequence reactances of a 15 MVA, 11kV synchronous generators are 0.3 pu, 0.2 pu, and 0.1pu respectively. The generator is solidly earthed and is unloaded. A line-to-line fault occurs at the terminals of the generator. Determine the fault current.

Solution. The sequence network is shown in Fig. 9.2.8 :

Assume that phases *b* and *c* are faulted.

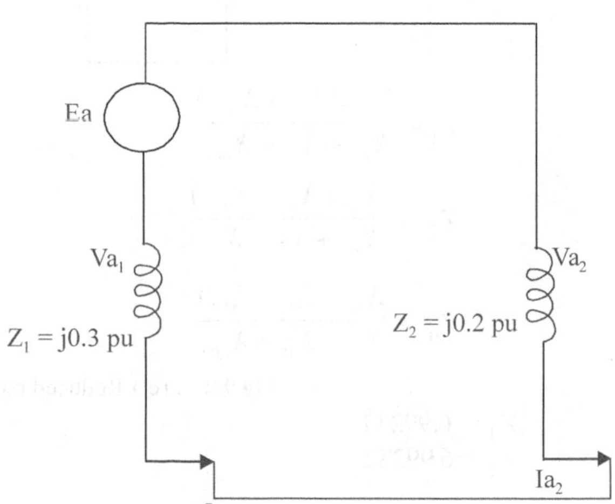

Fig.9.2.8. *Sequence network*

Let $E_a = 1\angle 0°$.pu

For the line-line fault,

$I_a = 0$

$I_{a1} = -I_{a2} = 1\angle 0° / (Z_1 + Z_2)$.

$Z_1 = 0.3i$

$Z_2 = 0.2i$

$I_{a1} = -I_{a2} = -2i$

The fault current is $I_b = I_{b0} + I_{b1} + I_{b2} = 0 + a^2.I_{a1} + a.I_{a2}$.

$a = -0.5 + 0.866i$

$a^2 = -0.5 - 0.866i$

$I_b = -3.464$ pu

Base MVA = 15

Base kV(L – L) = 11

Base current = 787.296 A

Fault current = – 2727.19 A

Prob.9.2.9. To find fault current & voltages for a DLG fault at the terminals of an unloaded generator

A double line to earth fault occurs at the unloaded generator terminals in Prob.8.2.8. Calculate the fault current and the line voltages

Solution.

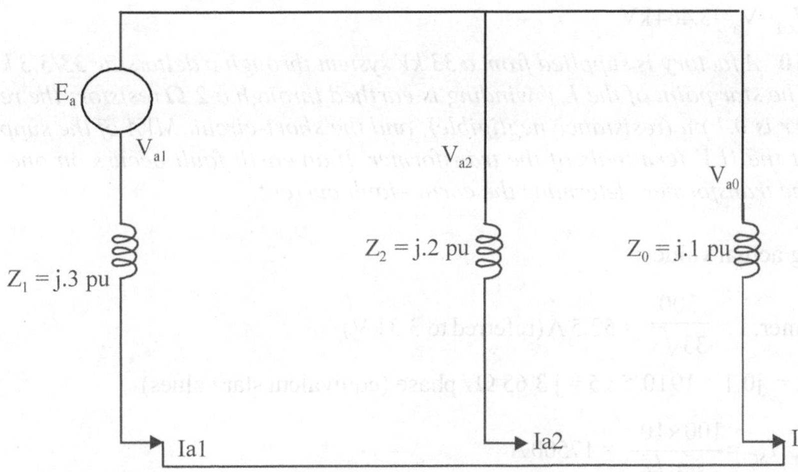

Fig.9.2.9. *Sequence network*

Assume that phases b and c are faulted to earth.

Let, $E_a = 1\angle 0°$.pu

$E_{a1} = 1$

$Z_1 = 0.3i$

$Z_2 = 0.2i$

$Z_0 = 0.1i$

$$I_{a1} = \frac{E_a}{Z_1 + \left(\frac{Z_0 Z_2}{Z_0 + Z_2}\right)} = -2.727i \text{ pu}$$

$V_{a1} = E_a - I_{a1}Z_1 = \quad = 0.181$ pu

$V_{a2} = V_{a1} = 0.181$ pu

$V_{a0} = V_{a1} = 0.181$ pu

$I_{a2} = V_{a2}/Z_2 = -0.909i$ pu

$I_{a0} = V_{a0}/Z_0 = -1.818i$ pu

The fault current to earth is $I_b + Ic$.

$I_b + I_c = (I_{a0} + a^2 I_{a1} + a.I_{a0}) + (I_{a0} + a.I_{a1} + a^2.I_{a2})$

$= 2.I_{a0} + (a + a^2)(I_{a1} + I_{a2})$

$= 2I_{a0} - (I_{a1} + I_{a2})$

Since $I_a = 0$, we may write $I_a = I_{a0} + I_{a1} + I_{a2} = 0$

or, $-(I_{a1} + I_{a2}) = I_{a0}$.

Therefore, fault current $= I_b + I_c = 3\ I_{a0} = -5.454\ i$ pu

Since Base current in Prob.9.2.8 = 787.3 A

Fault current = – 4294.363i

The line voltages are calculated as follows:

$V_a = 3V_{a1} = 0.545$ pu

$V_b = V_c = 0$

$V_{ab} = V_a = 3.464$ kV $(= 0.545 \times 11\sqrt{3})$

$V_{bc} = 0$

$V_{ca} = V_a = 3.464$ kV

Problem 9.2.10 *A factory is supplied from a 33 kV system through a delta/star 33/3.3 kV, 300 kVA transformer. The star point of the L.V winding is earthed through a 2 Ω resistor; the reactance of the transformer is 0.1 pu (resistance negligible), and the short-circuit MVA of the supply network is 100 MVA at the H.V. terminals of the transformer. If an earth fault occurs on one of the L.V. terminals of the transformer, determine the earth –fault current.*

Solution.

(*i*) Using actual values:

Transformer: $I = \dfrac{300}{33\sqrt{3}} = 52.5$ A (referred to 3.3 kV)

Hence, X = j0.1 × 1910/52.5 = j 3.65 Ω / phase (equivalent star values)

Network: $I_{SC} = \dfrac{100 \times 10^3}{3.3\sqrt{3}} = 17500$ A

Hence X= j 1910/17500 = j0.109 Ω / phase

Hence total impedance up to fault is

$Z_+ = j3.65 + j\,0.109 = j3.76 = Z^-$

$Zo = 3 \times 2 + j\,3.65$

$Z_+ + Z- + Z_o = 6 + j11.7 = 12.7$ (angle 60.7^0) Ω

Hence fault current $= \dfrac{3E}{Z_+ + Z_1 - Z_0} = \dfrac{3 \times 1910}{12.73} = 450$ A

(ii) Using pu values (base 300 MVA):

Base current $= \dfrac{300}{3.3 \times \sqrt{3}} = 52.5$ A

$$\text{Network impedance} = \frac{300}{100 \times 10^3} = j0.003 \text{ pu}$$

$$\text{Earthing resistor} = \frac{2 \times 52.5}{1910} = 0.055 \text{ pu}$$

$$Z_+ = Z- = j\,(0.1 + 0.003) = j0.103 \text{ pu}$$

$$Z_o = 3 \times 0.055 + j0.1 = 0.165 + j0.1 \text{ pu}$$

$$Z_+ + Z- + Z_o = 0.165 + j0.306 = 0.348\,(\angle\, 60.7^0) \text{ pu}$$

$$\text{Fault current} = 3\, I_{base} / (Z_+ + Z- + Z_o)0 \text{ pu} = 3 \times 52.5/0.348 = 450\text{A}$$

(Note the factor 3 that occurs in both cases in finding Z_o for the earthing resistor (all three zero-sequence current components pass through the single resistor) and also in the pu method the final factor 3 since fault current = $3I_+$.)

CHAPTER 10

PROTECTION

10.1 Introduction

It is required to generate sufficient electrical power at the most suitable locality, transmit it to the load centres at the most economic and ecological price giving due regard to security, continuity and reliability. Interconnected operation of a power system leads to economic power supplies. However, it presents stability problems and threatens the reliability & continuity of supply. The timely isolation of faulty sections of the system in the interest of stable operation cannot be overemphasized.

Students should be aware of the equipment for protecting plant against overcurrents and overvoltages, particularly the former. Knowledge of the characteristics of the various types of relay and other devices is essential. Soild–state electronic equipment is widely used and students should keep well abreast of developments in this direction.

10.1.1 Components of protection schemes

Identify the following components in a simple protection scheme :

1. Sensors for monitoring the operating conditions

2. Relay system to process and decide if isolation is desired and if so to send trip signals to the circuit breaker

3. circuit breakers to effect actual isolation.

The efficiency of a protection scheme depends upon its ability to

- correctly identify the faulty section.
- isolate it in minimum time.
- avoid false tripping.

10.1.2 Unit & Non-Unit type of protection

Discuss the basic difference between unit & non-unit type of protection. Give examples.

Distinguish between primary & back-up protection.

Stress the need for duplication of protection using 3-zone protection scheme.

10.1.3 Classification of relays

Based upon

1. principle of operation

2. speed of operation

3. derivation of relaying signals

4. functions performed

5. nature of actuating quantity (current, voltage, power, reverse power, etc.}

6. number of quantities measured and compared

7. method by which a relay acts upon the circuits protected-Direct/Indirect acting

- Distinguish between time overcurrent, directional and differential relays
- Explain

(*i*) Back-up protection

(*ii*) Directional feature of protection & its application

(*iii*) Negative sequence relay

• Explain the principle of operation of the induction relay. How are time & current settings provided ?

10.1.4 Functional requirements of a relay

1. Reliability
2. Sensitivity
3. Discrimination
4. Speed of operation

What are internal & through faults?

10.1.5 Methods of discrimination

1. Discrimination to fault location-
- By time
- By current magnitude
- By time and direction
- By distance measurement
- By time and current magnitude
- By current & voltage balance
- By direction of power
- By phase comparison

2. Discrimination to type of fault
- By sequence networks

3. Discrimination to location & type of fault

10.1.6 Basic structure of a relaying scheme

The blocks of a protection scheme should be explained.

Explain the
- Quantities measured by relays.
- Measuring element: means of measurement, CTs, PTS, and their characteristics.
- Comparing and control elements.

State the functional requirements of a relay.

Explain the requirement of discrimination, and methods employed for achieving proper discrimination.

Explain the following terms :
- Reset
- Pick-up
- Type of contacts
- Over-reach
- Definite, inverse, and very inverse characteristics of a relay

10.1.7 Power system protection

Principles of differential protection

- Circulating current systems
  Role of stabilizing resistor and biasing
- Balanced voltage systems
  Examples of differential relays

Principles of distance protection

3-zone protection, requirements of zonetime settings, schemes of 3-zone protection, relay characteristics in the impedance plane

Protection of system components

A. Generator protection for
- External faults, thermal overloading, unbalanced loading

- Stator protection for L-L,L-G, and inter-turn faults
- Rotor protection
- Reverse power protection

B. Transformer protection

- Differential protection
- Winding protection
- Buchholz relay
- Over-current protection

C. Busbar protection

1. Explain the methods & type of earth fault protection
2. Explain the different polar characteristics of distance protection indicating positive & negative torque regions
3. Explain the method of 3-zone protection and method of providing it.

10.1.8 Current and Voltage Transformers

Sources of error, effect of saturation, basic construction, especially for H.V units (including capacitor voltage transformers)

10.1.9 Relays

Attracted armature, induction, balanced beam, time-delay and directional relays. Simple equivalent transistor and thyristor circuits. Most of the above have a single input; double-input realays, one input operating and he other restraining are often necessary, *e.g.*, for distance protection, and can have their characteristic plotted on a complex R-X plain as shown in Fig. 10.1.9

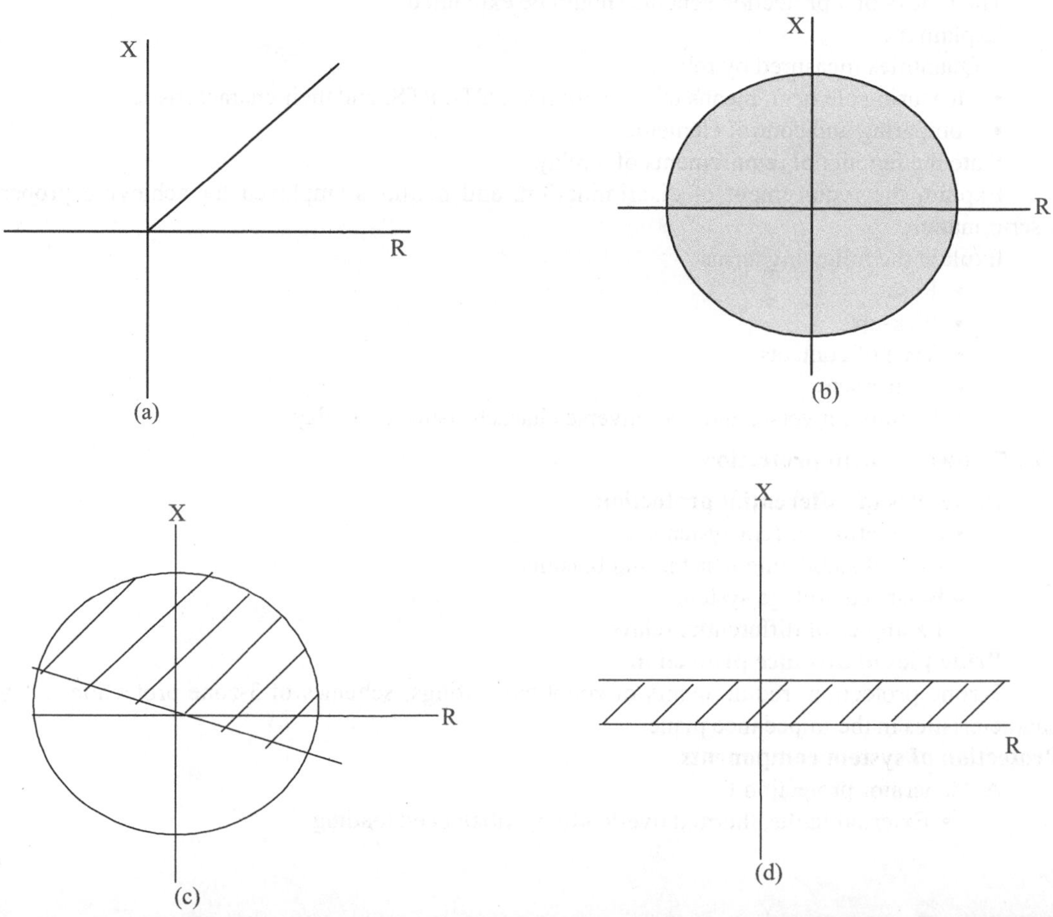

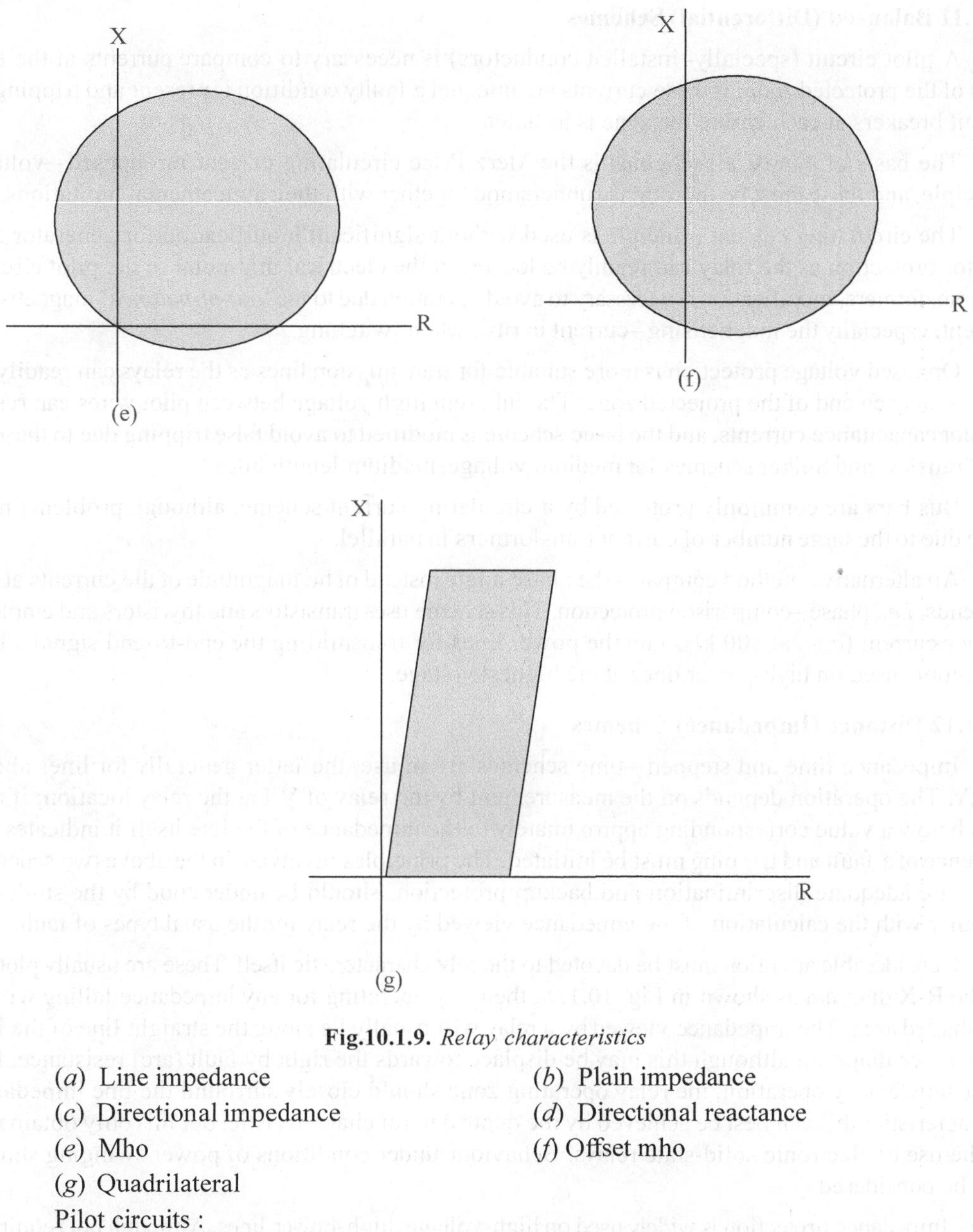

Fig.10.1.9. *Relay characteristics*

(*a*) Line impedance (*b*) Plain impedance
(*c*) Directional impedance (*d*) Directional reactance
(*e*) Mho (*f*) Offset mho
(*g*) Quadrilateral

Pilot circuits :

To transmit signals from end to end of a line specially installed conductors, Post-office circuits, power lines with carrier currents or microwave links can be used.

10.1.10 Overcurrent and earth leakage schemes

These are largely confined to circuits at 11 kV and below. However, they are sometimes used at 33 kV and for backup protection for other schemes of higher voltage lines. The combined use of overcurrent and directional relays, together with approximate time lags, achieves discrimination, *i.e.* selections and tripping out of only the faulty plant. Students should be able to determine appropriate settings for given ring-main or other simple networks and should be familiar with the methods of connecting directional relays to give adequate torque under all likely fault conditions.

10.1.11 Balanced (Differential) Schemes

A pilot circuit (specially- installed conductors) is necessary to compare currents at the two ends of the protected zone; if these currents are unequal a faulty condition is present and tripping of circuit breakers at each end of the zone is initiated.

The basis of almost all schemes is the Merz Price circulating current or opposed –voltage principle, and these must be thoroughly understood together with their fundamental limitations.

The circulating current principle is used without significant modification for generator and reactor protection as the relay can readily be located at the electrical mid-point of the pilot circuit. For transformers, modification is necessary to avoid operation due to the '*out-of-balance*' magnetising current, especially the magnetising –current in rush when switching.

Opposed voltage protection is more suitable for transmission lines as the relays can readily be located at each end of the projected zone. The inherent high voltage between pilot wires can result in pilot capacitance currents, and the basic scheme is modified to avoid false tripping due to these in the Translay and Solkar schemes for medium voltage, medium length lines.

Bus bars are commonly protected by a circulating current scheme, although problems may arise due to the large number of current transformers in parallel.

An alternative method compares the phase angle instead of he magnitude of the currents at the two ends, *i.e.*, phase –comparison protection. This scheme uses transistors and thyristors and employs carrier current (*e.g.*, at 500 kHz) on the power lines for transmitting the end-to-end signals. It is commonly used on high-power lines at the highest voltage.

10.1.12 Distance (Impedance) Schemes

Impedance time and stepped –time schemes are in use, the latter generally for lines above 33 kV. The operation depends on the measurement by the relay of V/I at the relay location; if this falls below a value corresponding approximately to the impedance of the line itself it indicates the presence of a fault and tripping must be initiated. The principles involved in the above two schemes to secure adequate discrimination and backup protection should be understood by the students, together with the calculation of the impedance viewed by the relay for the usual types of fault.

Considerable attention must be devoted to the rely characteristic itself. These are usually plotted on the R-X diagram as shown in Fig. 10.1.9., the relay operating for any impedance falling within the shaded area. The impedance viewed by a relay will usually lie along the straight line of the line impedance diagram, although this may be displace towards the right by fault (arc) resistance. For most satisfactory operation, the relay operating zone should closely surround the line impedance characteristic; this can best be achieved by the quadrilateral characteristic, but this only obtainable by the use of electronic solid-state relays. Behaviour under conditions of power swinging should also be considered.

Impedance protection is widely used on high-voltage, high-power lines. Although not requiring any pilot circuits, carrier –current signals over the power lies are sometimes used to intertrip the circuit breaker at the remote end simultaneously with that at the sending end, thereby eliminating the time lag that is otherwise necessarily introduced to secure discrimination.

10.1.13 Miscellaneous Schemes

The most important is negative sequence protection which detects the presence of the negative sequence current and hence the presence of a fault.

10.1.14 Overvoltage Protection

Protection against overvoltages may be by

(*i*) preventing the occurrence of overvoltage on the line
(*ii*) preventing the overvoltage reaching vulnerable plant, and
(*iii*) preventing circuit outages due to flashovers caused by such overvoltages.

The behaviour of a lightning stroke as it approaches a line, and the protection afforded by one or more earth wires , should be studied; also the nature of a back –flashover and the need for low tower-footing resistances. Switching surges can be minimised to some extent by resistance switching.

Vulnerable plants, particularly transformers, can be protected by surge diverters, *i.e.*, rod gaps or lightning arresters (non-linear resistance type) which divert the surge to earth before it reaches the plant being protected. The lightning arrester, although more expensive has the advantage of inherently clearing the power-follow current that succeeds the transient discharge, whereas with rod gap this has to be cleared by a circuit –breaker operation.

Students should be familiar with the principle of '*insulation co-ordination*', *i.e.*, matching the current /time breakdown characteristics of the protective devices with that of the transformer insulation.

Outages due to transient flashovers that do not cause permanent damage can be prevented by the use of arc-suppression coils or auto-reclose circuit breakers.

10.1.15 CONTROL & RELAY PANELS [IS: 8623 – 1993]

ROUTINE TESTS

S. NO.	*Routine test*	*Purpose*
1.	Wiring, electrical operation	To inspect the ASSEMBLY including wiring & function.
2.	Protective measures	To check protective measures by application of control voltage and current injection at secondary.
3.	Dielectric test at 2.5 kV for 1 min.	To verify withstanding high voltage test
4.	• Insulation resistance • Verification of items as per approved drawing/ Bills of materials • Verification of routine test reports of major bought out items, like relays, meters, etc.	• To verify insulation provided on conductive parts • To verify the quantities of required material in the panel as per approved drawing/B.O.M. • To verify routine test certificates of relays, meters etc.

Notes : **1.** Routine tests are to be considered as Acceptance Tests.
2. The routine test certificates of manufacturer of all bought out items are to be verified.

OBJECTIVE TYPE QUESTIONS

1. Which of the following relays gives a fore-warning of a possible major fault ?
(*a*) Buchholz relay
(*b*) Static Relay (*c*) Thermal relay
Ans. (*a*)

2. Which of the following conditions actuate a negative sequence relay ?
(*a*) L–L fault
(*b*) Overvoltages in the system (*c*) Reverse power flow
Ans. (*a*)

3. A relay is said to have an overreach if
(*a*) it can extend protection over a greater line length
(*b*) the measured fault impedance is greater than the actual one
(*c*) the measured fault impedance is less than the actual one
Ans. (*a*)

4. A zero sequence relay responds to
   (*a*) L– G fault
   (*b*) L– L fault (c̲) 3 – phase fault
   Ans. (*a*)
5. A typical zone 1 fault clearing time in a 3-zone distance protection scheme is about
   (*a*) 20 ms
   (*b*) 80 ms (*c*) 200ms
   Ans. (*b*)
6. A typical zone 2 reach of a 3-zone protection scheme is
   (*a*) 50%
   (*b*) 80% (*c*) 150%
   Ans: (*c*)

PROBLEMS WITH SOLUTIONS

Prob. 10.2.1. Overcurrent relay: Time of operation

Explain how would you use an overcurrent characteristic to determine its time of operation, given the following information:
Relay specification: 1A, 3 sec
Plug setting: 125%
Time multiplier setting: 0.6
Current transformer ratio: 400/1
Fault current: 4000 A

Solution.

C.T Ratio = 400
Fault current = 4000 A
Pilot wire current = Fault current / C.T Ratio = Relay coil current
Pilot wire current = 10 A
Plug setting = 1.25
Normal relay coil current = Rated current × Plug setting = 1× 1.25 = 1.25 A
Plug setting multiplier = Pilot wire current/Plug setting
= 8

i.e., Relay fault current as a multiple of plug setting = $\frac{10}{1.25}$

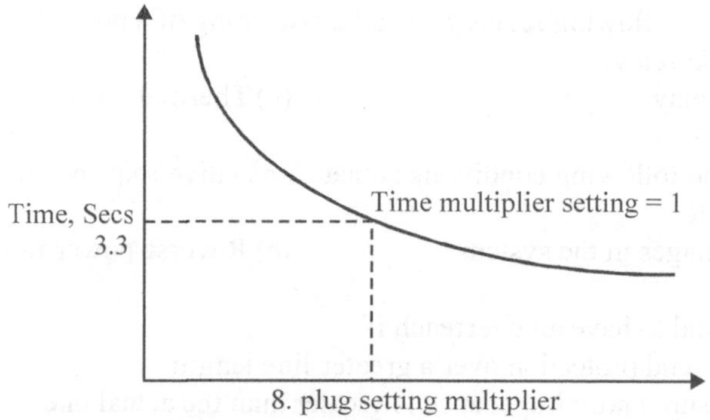

Fig. 10.2.1. *Relay curve*

From the relay curve, the time of operation is 3.3 for a time setting of 1
Operating time for TMS = 1 is 3.3 sec
Given, TMS = 0.6
Actual operating time = operating time for TMS = 1 × given TMS 1.98 sec **Answer**

Prob. 10.2.2. Transformer-Differential Protection

A 3-phase delta-star-connected 30 MVA 33/11 kV transformer is protected by a differential relay. Calculate the relay current setting for faults drawing up to 200 % of the rated current. The CT current ratio on the primary side is 500:5 and that on the secondary side is 2000:5.

Solution.

MVA	Rating	= 30 MVA
kV_1	High side	= 33 kV
kV2	Low side	= 11 kV

I_p Primary line current $= \dfrac{MVA \times 1000}{\sqrt{3}\, K.V_1}$

$I_p = 525.486$ A

I_s Secondary line current $= 3 \times I_p = 1576.458$A

CT1 ratio primary side = 100

CT2 ratio primary side = 400

C.T current on primary side, $I_1 = I_p$/CT1 ratio = 5.254A

C.T current on sec.side, $I_2 = 1.73 \times$ Is/CT2 rating = 6.826A

Relay current at 200 % of rated current $= 2 \times (I_2 - I_1) = 3.142$ A **Answer**

Prob. 10.2.3. Transformer- differential protection-CT Ratios

A 3-phase delta-star connected 30MVA, 132/33 kV transformer is protected by current transformers. Determine the CT ratios for differential protection such that the circulating current (through the transformer delta) does not exceed 5 A.

Solution. HV L-L = 132 kV

LV L-L = 33 kV

MVA = 30 MVA

$$\text{I delta, Line value} = \frac{MVA \times 1000}{\sqrt{3}\, HV}$$

= 131.371 A

$$\text{I star, Line value} = \frac{MVA \times 1000}{\sqrt{3}\, LV}$$

= 525.486

I_c, Circulating current = 5 A

The CTs on the HV side are connected in star. Therefore,

CT ratio HV side = 105.097 **Answer**

CTs are connected in delta on the LV side. Therefore,

$$\text{CT ratio LV side} = 131.37\left(\frac{5}{\sqrt{3}}\right) = 379.687$$ **Answer**

Prob. 10.2.4. Circuit breaker ratings

A circuit breaker is rated at 1000 A, 1200 MVA, 33 kV, 3 sec, 3-phase oil-circuit breaker.

(*i*) *the rated normal current*
(*ii*) *breaking capacity*
(*iii*) *rated symmetrical breaking current*
(*iv*) *rated making capacity*
(*v*) *short-time rating*
(*vi*) *rated service voltage*

Solution. I = 1000A

MVA = 1200

kV = 33

(*i*) = 1000 A **Answer**

(*ii*) = 1200 MVA **Answer**

(*iii*) $= \dfrac{MVA \times 1000}{\sqrt{3}\,KV} = 21019.44$ A (rms) **Answer**

(*iv*) Maximum value of asymmetrical current = 2.55 × Rated symmetrical Breaking current = 53599.58 A **Answer**

(*v*) Same as rated symmetrical breaking current = 21019.44 A for 3 seconds. Answer

(*vi*) 33 kV **Answer**

CHAPTER 11

STEADY-STATE STABILITY

11.1 Introduction

Power system stability is a term applied to alternating current electric power systems, denoting a condition in which the various synchronous machines of the system remain in synchronism, or "*in step*" with each other. Conversely, instability denotes a condition involving loss of synchronism, or falling "*out of step*".

The AIEE standard definition of stability is as follows, **"Stability when used with reference to a power system, is that attribute of the system, or apart of the system which enables it to develop restoring forces between the elements thereof, equal to or greater than the disturbing forces so as to restore a state of equilibrium between the elements".**

11.1.1 Small and large perturbations

A power system is subjected to a variety of disturbances. These are classified into two categories:

- Small perturbations
- Large perturbations

Small Perturbations

Perturbations are characterized as small if the changes in system states are small due to these perturbations. The magnitude of perturbation is small enough o allow the use of linearized state equations obtained by linearizing the nonlinear differential equations around the operating point for studying he dynamics of the system. Random changes in load which occur in the system continuously are an example of small perturbation. Stability problem associated with small perturbation is known as dynamic stability.

Large perturbations

A power system may be subjected to large perturbations such as :

- Occurrence of faults on the line
- Loss of large generating units
- Loss of major transmission facilities
- Loss of large loads

The stability problem associated with large perturbations is known as transient stability.

Exercises

1. Derive and explain the power -angle characteristic of
   - Machine infinite-bus system
   - Two machine system
2. Discuss the concepts of stable & unstable equilibrium points.
3. What is meant by steady state & transient stability of a system ?

11.1.2 Voltage Stability

A basic introduction to the static voltage instability problem may be provided, by using a simple generator connected to a load through two parallel transmission lines [Fig.11.1.2 (a)]. The student can plot PV, QV, or PQ curves for different values of line reactance, generator terminal voltage, and power factor.

Parameters can be changed, if necessary. Different plots will provide the student with the *basic* material needed to understand PV, QV, or PQ curves [Figs.11.1.2 (b, c, d)]. The concept of maximum power transfer can be studied.

Sample questions are given below :

1. The student is asked to calculate the reactive power required to maintain certain voltage V_r at the load bus for a given P_r, V_s, and X_l.
2. The student is asked to calculate the reduction in maximum power transfer with one line out.
3. In another question, the student may be asked to display dynamically the PV trajectories followed by the system as the demand increases.

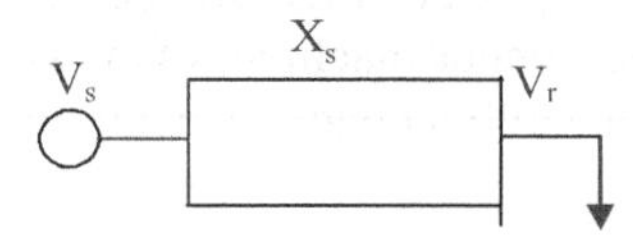

Fig.11.1.2. (*a*) *A simple transmission system*

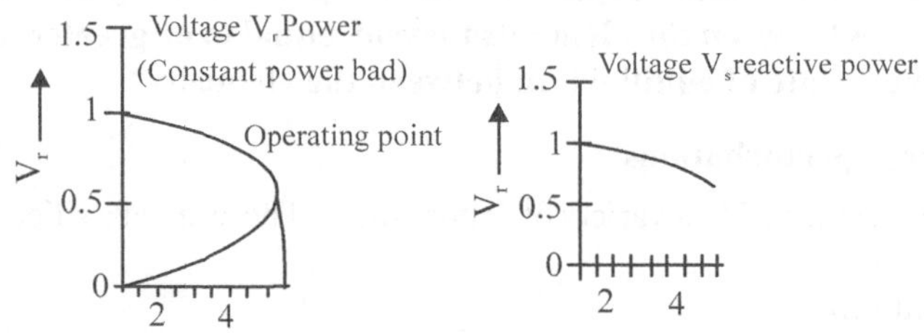

Fig. 11.1.2. (*b*) *P.V. Curve*

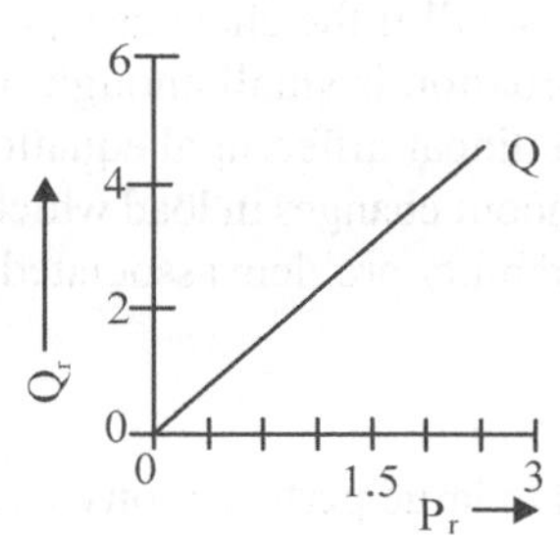

Fig. 11.1.2. (*d*) *P.Q Curve*

11.1.3 Power Transfer

The power that can be transmitted over a transmission circuit is limited by thermal considerations, economics, steady-state power limits and transient power limits.

Thermal limits for underground cables are amenable to calculations, but for overhead lines experimental data must be relied on.

The surge impedance load, $\dfrac{V_{ph}^{\ 2}}{Z_o}$ watts/phase, gives minimum loss and voltage drop and is desirable or essential for long lines.

11.1.4 Steady-State power limits

For given voltages at the ends, the power transmitted over the line

$$P=\frac{V_s V_r \cos(\beta-\delta)}{B}-\frac{A V_r^{\,2}\cos(\beta-\alpha)}{B} \text{ watts/phase}$$

where V_s, V_r are phase voltages and B = *B (angle β)*, A = *A (angle α)*.
From this the maximum power is

$$P_M = \frac{V_s V_r}{B} - \frac{A V_r^2 \cos(\beta - \alpha)}{B}$$

This reduces to

$$P = \frac{V_s V_r}{X} \sin\delta \text{ and } \frac{V_s V_r}{X}, \text{ respectively}$$

for a short line with negligible resistance.

OBJECTIVE TYPE QUESTIONS

1. The power angle characteristic of a machine-infinite bus system is P = 2 sin δ (pu). The initial operating angle is 60°., inertia constant H = 5 sec. System frequency is 50 Hz. The angular frequency of oscillation following small perturbation will be

(*a*) $\sqrt{31.4}$ rad/sec

(*b*) $\sqrt{15.7}$ rad/sec

(*c*) $\sqrt{62.8}$ rad/sec

Ans. (*a*)

2. In an interconnected power system, the frequency of electro-mechanical modes of oscillation lies in the range
(*a*) 0.5-2.5 Hz
(*b*) 1-10 Hz
(*c*) 30-60 Hz
Ans. (*a*)

3. If the sending end and receiving end voltages for a 3-phase transmission line are each 33 kV (line), and if the reactance of the line is 13 Ω / phase, the maximum power transmitted per phase will be
(*a*) 60 MW (*b*) 30 MW
(*c*) 29 MW (*d*) 28 MW
Ans. (*d*)

4. The torque angle corresponding to the steady-state stability limit of a salient-pole alternator is
(*a*) Less than 90°.
(*b*) Greater than 90°.
(*c*) Equal to 90°.
Ans. (*a*)

5. When the alternator stalls (near the stability limit) the armature current is
(*a*) Less than the rated value
(*b*) Greater than the rated value
(*c*) Equal to the rated value.
Ans. (*a*)

PROBLEMS WITH SOLUTIONS

Prob.11.2.1. Frequency of rotor oscillations for small disturbance

A synchronous generator of reactance 1.5 pu is connected to an infinite bus bar (V = 1 pu) through a line and transformer of total reactance 0.5 pu. The no-load voltage of the generator is 1.2 pu and the inertia constant H = 3.1416 MW-sec per MVA. Resistance & damping may be neglected. Calculate the frequency of the oscillations set up when the operates at load angle of 60° and is subjected to a small disturbance.

Solution. X_g = 1.5 pu

Xline & Transformer = 0.5 pu

X Total = 2 pu

δ_0 = 60°

E = 1.2 pu

V = 1 pu

$dp/d\delta$ = EV cosδ/X = 0.300

G = 1 MVA

H = 3.1416

f = 50 Hz

M = GH/(πf) = 0.02

$$p_{1,2} \text{ roots} = +\sqrt{\left(\frac{\delta P}{d\delta_0}\right)/M}$$

$= \pm$ j 3.874 rad/s

f = 0.616 Hz **Answer**

Prob.11.2.2 Steady -state stability limit of a generator connected to an infinite bus via parallel lines

A 3-phase generator transformer set is connected to an infinite bus bar via duplicate lines. and the steady-state stability limit is 200 MW. This limit is reduced to 70 MW during a 3-phase short-circuit midway along one-line. Calculate the pu reactance on a base of 100 MVA of the generator-transformer and of each line, and the power limit when the faulty line is isolated.

What reasonable assumptions must be made regarding the system e.m.f'.s?

Solution.

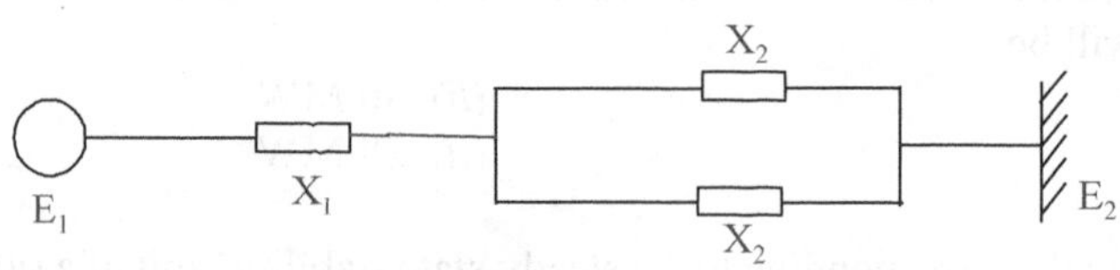

Fig. 11.2.2(a). *Two lines in parallel*

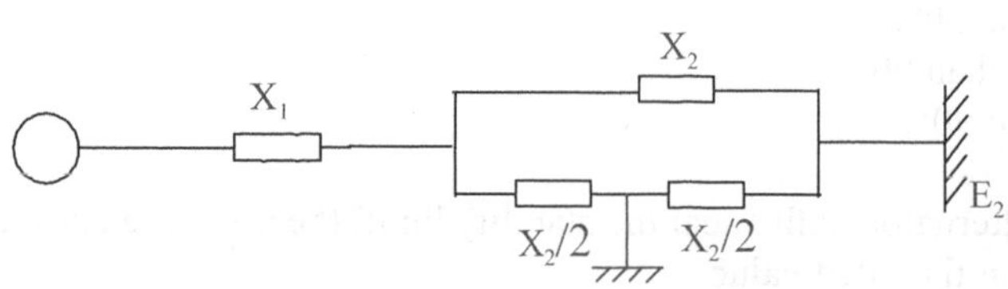

Fig. 11.2.2(b). *One line faulted*

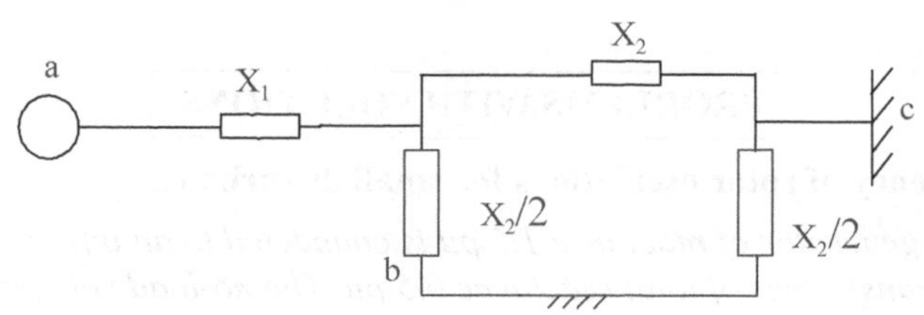

Fig. 11.2.2(c). *Star connection, abc*

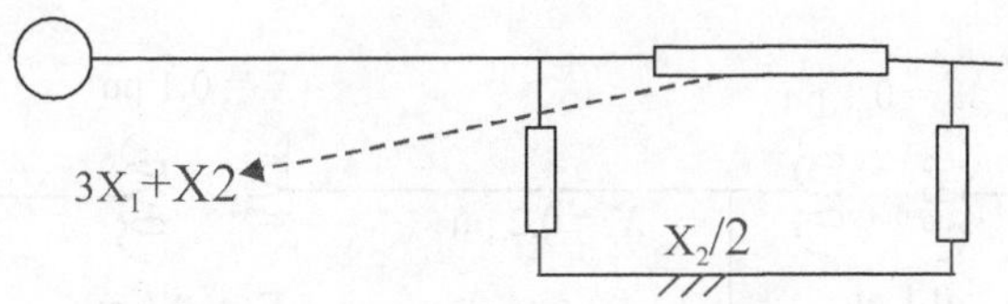

Fig. 11.2.2(d). *Star connection converted to delta*

$E_1 = 1$ pu
$E_2 = 1$ pu
Unfaulted lines -Reactance between E_1 & $E_2 = X_1 + (X_2.2)$
One line faulted-Reactance between E_1 & $E_2 = 3X_1 + X_2$
Steady state stability – Unfaulted lines
$P_{max}1 = 200$ MW
= 2 pu on 100 MVA base
Steady state stability -faulted line
$P_{max}2 = 70$ MW on 100 MVA base
= 0.7 pu
Therefore,

$$X_1 + \frac{X_2}{2} = \frac{E_1 E_2}{P_{max1}} = 0.5 \quad \text{. . . (1)}$$

$3X_1 + X_2 = E_1E_2/P_{max2} = 1.428$ pu . . . (2)
Rewrite Eqs. (1) & (2)
$3X_1 + 1.5X_2 = 1.5$ pu . . . (3)
$3X_1 + X_2 = 1.428$ pu . . . (4)
Subtract (4) from (3) :
$0.5X_2 = 0.071$
$X_2 = 0.142$ pu **Answer**
From Eq.4, we get
$X_1 = 0.428$ pu **Answer**
Power limit when the faulty line is isolated:

$$P_{max}3 = \frac{E_1 E_2}{X_1 + X_2}$$

= 1.749 pu
= 174.999 MW **Answer**

Prob.11.2.3 Steady-state stability of a system with & without a shunt reactor

Calculate the steady-state power limit of the system (shown in Fig.11.2.3 (a)) in pu and in MW with and without the reactor connected.

All pu reactances are expressed on 10MVA Base

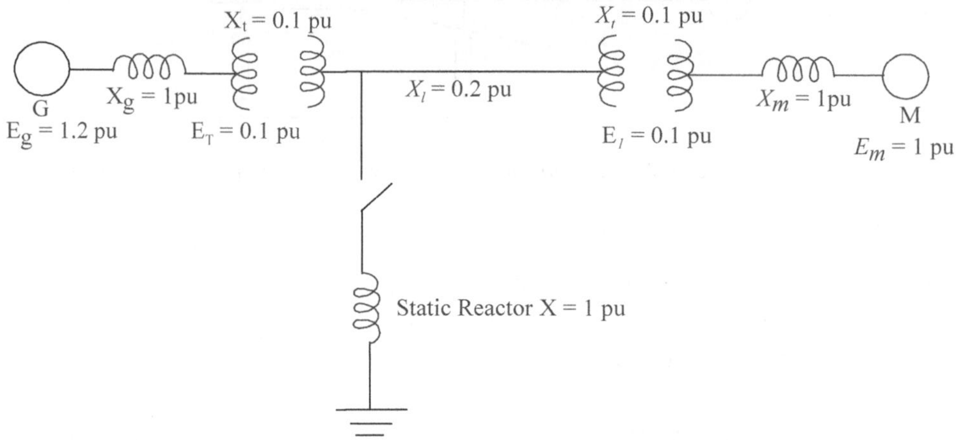

Fig.11.2.3. *(a) Power system*

Solution: Without Reactor:

$$P_{max} = E_g.E_m/(X_g + X_T + X_l + X_T + X_m) = 0.5 \text{ pu} = 5 \text{ MW}$$ **Answer**

$$= 5 \text{ MW}$$ **Answer**

With Reactor:

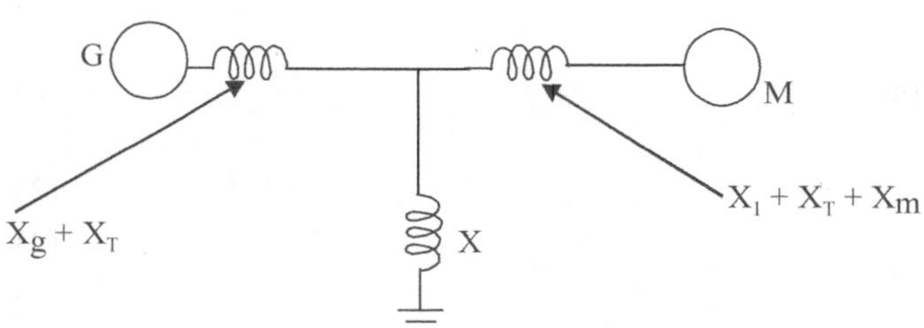

Fig.11.2.3 (b). *T-equivalent circuit*

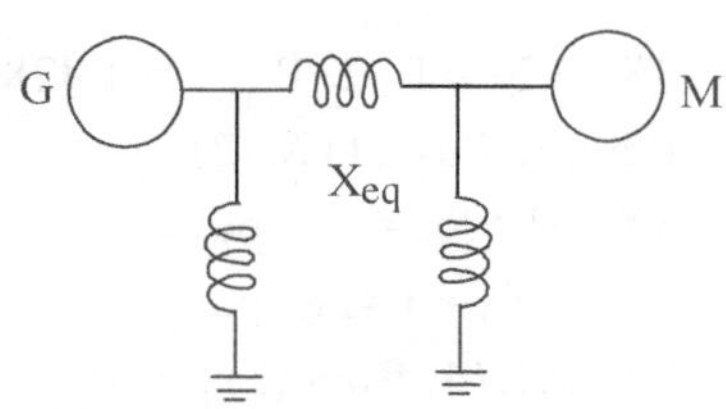

Fig.11.2.3 (c) PI- equivalent circuit

$$X_{eq} = (X_g + X_T) + (X_l + X_T + X_m) + (X_g + X_T).(X_l + X_T + X_m)/X = 3.83 \text{ pu}$$

$$P_{max} = \frac{E_g.E_M}{X_{eq}} = 0.313 \text{ p.a.} = 3.133 \text{ MW}$$ **Answer**

Prob. 11.2.4 Steady -state stability of a generator connected to an infinite bus via a line with a resistive load connected halfway-Example 1

Figs. 11.2.4 (a) and (b) show respectively a power system and its reactance diagram.

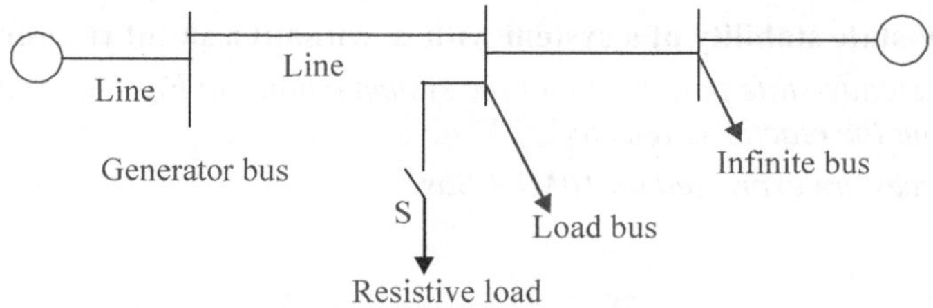

Fig.11.2.4. *(a) Power system*

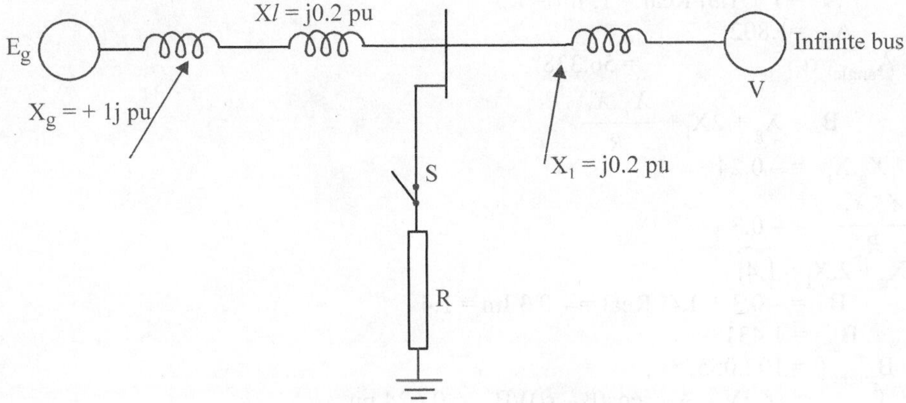

Fig.11.2.4 (*b*) *Equivalent circuit with S open*

Voltage of infinite bus is 1 pu.

With S open, the complex power delivered to the infinite bus is = 0.8 + 0.6 jpu

S is then closed through a load resistance, R= 0.8 pu.

At the same time generator excitation is kept constant and mechanical power input to the generator is increased to maintain the power input to the infinite bus at 0.8 pu.

Determine whether the steady-state stability limit is reached.

Solution. S open:

Voltage of infinite bus = V = 1 + 0j pu Vabs = 1

Current into infinite bus is denoted by I.

Power into infinite bus = 0.8 + 0.6j

Conjugate Power = 0.8 – 0.6j

Therefore, I = 0.8 – 0.6j

$E_g = V + I(X_g + X_l + X_l)$

$I(X_g + X_l + X_l) = 1.4j$

$I(X_g + X_l + X_l) = 0.84 + 1.12j$

$E_g = V + I(X_g + X_l + X_l) = 1.84 + 1.12j$

E_g magnitude = 2.154

E_g angle = 31.344°

S closed:

E_g remains constant, and the circuit is shown in Fig.11.2.4(c).

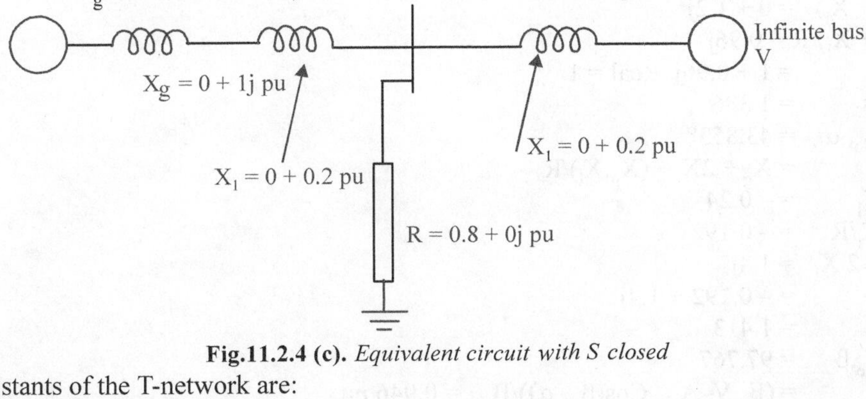

Fig.11.2.4 (c). *Equivalent circuit with S closed*

A, B constants of the T-network are:

$A = 1 + (X_g + X_l)/R$

$(X_g + X_l) = 1.2j$

$(X_g + X_l)/R = 1.5j$

$A = 1 + 1.5j$ Real = 1, Im = 1.5

$A_{abs} = 1.802$

$A_{angle}, \alpha = 56.338°$

$$B = X_g + 2X_l + \frac{X_g.X_l}{R}$$

$X_g.X_l = -0.24$

$$\frac{X_g.X_l}{R} = -0.3$$

$X_g + 2.X_l = 1.4j$

$B = -0.3 + 1.4j$ Real = – 0.3 Im = 1.4

$B_{abs} = 1.431$

$B_{angle}, \beta = 102.0552°$

$P_{max} = (E_g.V - A_{abs}\cos(\beta - \alpha))/B_{abs} = 0.624$ pu

The desired delivery of power of 0.8 pu to the infinite bus cannot be met. System is unstable

Prob.11.2.5 Steady -state stability of a generator connected to an infinite bus via a line with a resistive load connected halfway-Example 2

Determine whether delivery of power of 0.8 pu to the infinite bus of the system for Prob.11.4 is possible if the load resistance is 1.25 pu.

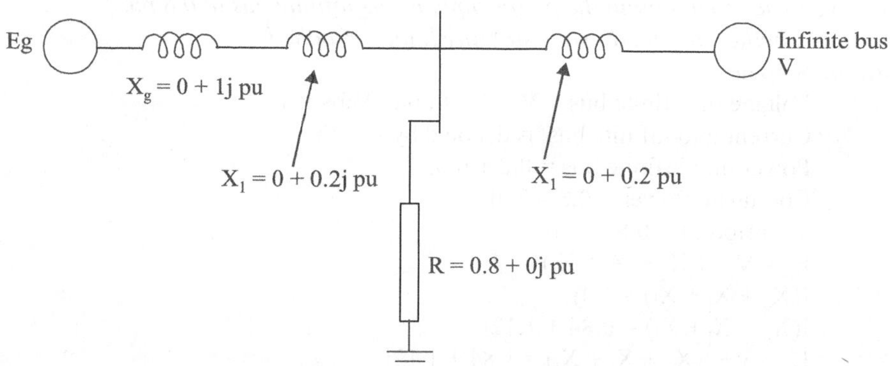

Fig.11.2.5. *Power system*

A, B constants of the T-network are:

$A = 1 + (X_g + X_l)/R$

$(X_g + X_l) = 0 + 1.2j$

$(X_g + X_l)/R = 0.96j$

$A = 1 + 0.96j$ Real = 1

Aabs = 1.386

$A_{angle}, \alpha = 43.853°$

$B = X_g + 2X_l + (X_g.X_l)/R$

$X_g.X_l = -0.24$

$X_g.X_l/R = -0.192$

$X_g + 2.X_l = 1.4j$

$B = -0.192 + 1.4j$

$B_{abs} = 1.413$

$B_{angle}, \beta = 97.767$

$P_{max} = (E_g.V - A_{abs}\cos(\beta - \alpha))/B_{abs} = 0.946$ pu

P_{max} is greater than 0.8 pu. Hence, power of 0.8 pu to the infinite bus can still be maintained when the load resistance is 1.25pu.

CHAPTER 12

TRANSIENT STABILITY

12.1 Introduction

Look up the websites

- http://www.iea.lth.se/~ielmatsl/ObjectStab/- a free Modelica Library for Power Systems, suitable for the simulation and analysis of transient and voltage stability in power systems.
- http://www.uwaterloo.ca/ -TEFTS-free software- transient stability program to study transient energy functions and voltage stability issues- includes example & a brief tutorial.

The behaviour of the system subsequent to a sudden load change, switch operation or fault must be understood and the determination of the subsequent swing curve fully studied. The simplified equation of motion for a generator [$M\ (d^2\delta/\Delta t^2) = P$] is normally adequate for use in the usual step-by-step method of calculation. The general procedure for handling multimachine problems by computer should be understood.

The basis for, the applications and limitations of, the equal-area criterion for stability should be understood

12.1.1 Swing Equation

The motion of a synchronous machine is governed by the laws of mechanics applicable to rotating bodies. Net accelerating torque acting on the rotor is equal to the product of the angular acceleration and moment of inertia, *i.e.*,

$$J.\ d^2\theta/dt^2 = T_a = T_m - T_e \qquad \ldots (1)$$

where J = combined moment of inertia of the synchronous generator & the prime mover,

T_a = Net accelerating torque

T_m = Shaft torque

Te = Electromagnetic torque

θ = angular position of the rotor with respect to the stationary axis.

Both T_m and T_e are taken as positive for generator action, *i.e.*' mechanical input & electrical output. They are negative for motor action, *i.e.*, electrical input & mechanical output.

Since the rotor speed relative to synchronous speed is of interest, it is more convenient to measure the angular position and angular velocity with respect to the synchronously rotating reference axis.

We define

$$\theta = \omega_s t + \delta \qquad \ldots (2)$$

where ω_s = synchronous speed in rad/sec

t = time, sec.

δ = displacement angle of the rotor with respect to reference axis rotating at synchronous speed, ω_s.

From equations (1) and (2), we get

$$J\frac{d^2\delta}{dt^2} = T_M - T_c \qquad \ldots (3)$$

Multiplying both sides of this equation by ω, we get the swing equation

$$\frac{Md^2\delta}{dt^2} = P_M - P_C \qquad \ldots (4)$$

where M = Jω is the angular momentum

$P_m = \omega T_m$ is the shaft power input

$P_e = \omega T_e$ is the electrical power output corrected for electrical losses.

The angular momentum M of a machine is not strictly constant because the angular velocity ω varies somewhat during the swings following disturbances, but M may be assumed constant since speed of the machine does not differ much from the synchronous speed unless the synchronism is actually lost. M evaluated at synchronous speed ω_s is known as the inertia constant of the machine.

12.1.2 Units of M

1. If P_m and P_e are expressed in watts, δ in radians and t in seconds, then M is in Joule-sec/radian.

2. If P_m and P_e are expressed in megawatts, δ in electrical degrees and t in seconds, then M is in MJs/electrical degree.

3. If P_m and P_e are expressed in pu, δ in electrical degrees and t in seconds, then M is in pu power-sec^2/electical degree *i.e.* M is in pu.

12.1.3 Inertia Constant H :

Inertia constant H is different from inertia constant M. For synchronous machine inertia constant H is frequently specified and is defined as the ratio of the stored K.E at rated speed to the rated apparent power of the machine, *i.e.*

H = Stored K.E in MJ at synchronous speed / machine rating in MVA ... (5)

Swing equation (4) reduces to the form

$$\left(\frac{2H}{\omega_s}\right)\frac{d^2\delta}{dt^2} = P_M - P_e$$

where P_m and P_e are pu powers, d and was should have consistent units

12.1.4 Linearization of swing equation

For small perturbations, the dynamic behaviour of the system can be studied by linearizing the swing equations around the nominal operating point. The liberalised swing equation is

$$\frac{d^2\Delta S}{dt^2} + \omega_s \cdot \frac{S}{24} \cdot \Delta S = 0$$

where Δδ= small change in nominal operating angle δ

S = Synchronizing power coefficient = $dP_e/d\delta$ (at $\delta = \delta_o$)

12.1.5 Angular frequency of undamped oscillations, ω_n

$$\omega_n = \sqrt{\frac{(\omega_s . S)}{2H}} \text{ elect. radians/sec}$$

Exercise-What is the range of undamped natural frequency of oscillation of a synchronous generator?

12.1.6 Equal area criterion (EAC) of stability

To determine whether a power system is stable after a disturbance, it is necessary in general, to plot and inspect the swing curves. If these curves show that the angle between any two machines tends to increase without limit, the system of course is unstable. If on the other hand, after all the disturbances including switching have occurred, the angles between the two machines of every possible pair reach maximum values and thereafter decrease, it is probable, although not certain, that the system is stable. Occasionally in a multi-machine system one of the machines may stay in step on the first swing and yet go out of step on the second swing because the other machines are in different positions and react differently on the first machines.

In a two-machine system, under the usual assumptions of constant input, and constant voltage behind transient reactance, the angle between the machines either increases indefinitely or else, after all disturbances have occurred, oscillates with constant amplitude. In other words, the machines either fall out of step on first swing or never. Under these conditions the observation that the machines come to rest with respect top each other may be taken as the proof that the system is stable. There is a simple graphical method of determining whether the machines come to rest with respect to each other. This method is known as the equal area criterion of stability. When this criterion is applicable its use wholly or partially eliminates the need of computing swing curves and thus saves considerable amount of computation time. It is applicable to any two-machine system for which the assumptions stated above may be made.

12.1.7 Application of EAC

Consider a machine infinite bus system. The swing equation of the system is

$M\, d^2\delta/dt^2 = P_m - P_e = P_a$

Multiplying both sides of the equation by (2/M) dδ/dt, we get

$2\, d\delta/dt\, .d^2\delta/dt^2 = (2Pa/M)\, d\delta/dt$ . . . (6)

Since $d\delta/dt.\, (d\delta/dt)^2 = 2\, d\delta/dt\, .d^2\delta/dt^2$

Eq. (6) reduces to

$d\delta/dt.\, (d\delta/dt)^2 \cdot (2P_a/M)\, d\delta/dt$

Next, multiply each side by dt, obtaining

$d\,[(d\delta/dt)^2] = (2P_a/M)d\delta$

Integrating this equation, we get

$$\left(\frac{d\delta}{dt}\right)^2 = \frac{2}{M}\int_{\delta_0}^{\delta} Pa.d\delta$$

or,

$$\frac{d\delta}{dt} = \sqrt{\frac{2}{M}\int_{\delta_0}^{\delta} Pa.d\delta}$$

When the machine comes to rest with respect to the infinite bus- a condition, which may be taken to indicate stability-requiring that

$$\int_{\delta_0}^{\delta_M} Pa.d\delta = 0$$

This integral may be integrated graphically as the area under a curve of P_a plotted against δ between limits δ_o, the initial angle, and δ_m, the final angle. Areas A_1 and A_2 may be interpreted in terms of kinetic energy gained /lost by the synchronous generator.

Exercise: Illustrate the application of equal area criterion by applying it (to the following two simple cases) for a synchronous generator connected to infinite bus through a double-circuit line.

- Sustained fault
- Line fault cleared after the lapse of a certain time by the simultaneous opening of the circuit breakers at both ends of the line.

12.1.8 Transient stability limit

Transient stability limit of a two-machine system is defined as the maximum power that can be transmitted from one machine to the other without loss of synchronism for a specified sudden, severe, unrepeated shock.

Exercises:

- Illustrate the concept of transient stability limit using equal area criterion of stability.
- Obtain an equivalent one-machine infinite bus system of a two finite machines system
- Obtain an expression for the critical clearing angle
- What is the effect of location and duration of fault on the transient stability limit?

12.1.9 Numerical technique for solution of swing equation

The transient stability analysis requires the solution of a system of coupled non-linear differential equations. In general, no analytical solution of these equations exists. However, techniques are available to obtain approximate solution of such differential equations by numerical methods and one must therefore resort to numerical computation techniques commonly known as digital simulation. Some of the commonly used numerical techniques for the solution of the swing equation are:

- Point by point method
- Euler's method
- Euler's modified method
- Runge-Kutta method, etc.

12.1.10 Point -by -Point Methods

Method 1

Point by Point solution, also known as step-by-step solution is the most widely used way of solving the swing equation. The following two steps are carried out alternately.

1. First, compute the angular position δ, and angular speed $d\delta/dt$ at the end of the time interval using the formal solution of the swing equation from the knowledge of the assumed value of the accelerating power and the values of δ and $d\delta/dt$ at the beginning of the interval
2. Then compute the accelerating power of each machine from the knowledge of the angular position at the end of the interval as computed in step 1.

There are two different point-by-point methods. Method 2 is more accurate compared to method 1.

Method 2

In this method the accelerating power during the interval is assumed constant at its value calculated for the middle of the interval.

The desired formula for computing the change in δ during the nth time interval is

$$\Delta\delta n = \Delta\delta\, n{-}1 + \frac{(\Delta t)^2}{M} P_a(n-1)$$

where,

$\Delta\delta n$ = change in angle during the nth time interval

$\Delta\delta n - 1$ = change in angle during the $(n-1)^{th}$ time interval

Δt = length of time interval

$P_a\,(n-1)$ = accelerating power at the beginning of the nth time interval

Due attention is given to the effects of discontinuities in the accelerating power P_a which occur, for example, when a fault is applied or removed or when any switching operation takes place. If such a discontinuity occurs at the beginning of an interval, then the average of the values of P_a before and after the discontinuity must be considered. Thus, in computing the increment of angle occurring during first interval after a fault is applied at $t = 0$, the above equation becomes:

$\Delta\delta_1 = [(\Delta t)^2/M]\, Pa_0+/2$

where Pa_0+ is the accelerating power, immediately after the occurrence of the fault.

If the fault is cleared at the beginning of the mth interval, then for this interval,

$P_a(m-1) = 0.5\,[P_a(m-1)^- + P_a(m-1)^+]$

where $P_a(m-1)^-$ is the accelerating power before clearing and $P_a(m-1)^+$ is that immediately after clearing the fault. If the discontinuity occurs at the middle of the interval, no special treatment is needed.

12.1.11 Improvement of power system stability

System parameters affecting stability :

- Synchronous machine parameters
- Transmission line parameters
- Circuit breaker & relay characteristics
- System layout
- Excitation system and governor characteristics
- Neutral grounding

Discrete Controls:

- Auto -reclosing of circuit breakers
- Single-pole switching
- Resistor breaking
- Generator dropping
- Load shedding
- Series capacitor switching
- Shunt reactor/capacitor switching
- Boosting power on D.C. links

Fast Valving

Exercise-Discuss the effect of variation of M and auto-reclosure on the transient stability limit.

OBJECTIVE TYPE QUESTIONS

1. The critical clearing angle of a given power system for a certain fault is
(a) proportional to the Inertia Constant M
(b) proportional to the Inertia Constant M
(c) independent of M.
Ans. (*c*)

2. Equal area criterion of stability is applicable to
(a) a machine infinite bus system only
(b) both to a machine-infinite bus and a two-machine system
(c) a multi-machine system.
Ans. (*b*)

3. In a two-machine power system, machine A delivers power to machine B. A 3-phase fault occurs at the terminals of machine A. Initial acceleration of machine A is
(a) positive (b) zero (c) negative
Ans. (*a*)

4. Which one of the following enhances the transient stability of a system the most
(a) proper choice of make and break capabilities of the circuit breakers
(b) installation of 3-pole auto-reclose circuit breakers
(c) installation of single pole auto-reclose circuit breakers
Ans. (*c*)

PROBLEMS WITH SOLUTIONS

Prob.12.2.1 Equal area criterion of stability- Example 1

A generator delivers 1 pu power to an infinite bus bar through a network. A fault occurs which reduces the maximum power transferable to 0.4 pu whereas before the fault this power was 1.8 pu and after clearance of the fault 1.3 pu. Calculate the critical clearing angle , sketch the power - angle curves and show the equality of the accelerating and decelerating areas.

Resistance may be neglected.

Solution.

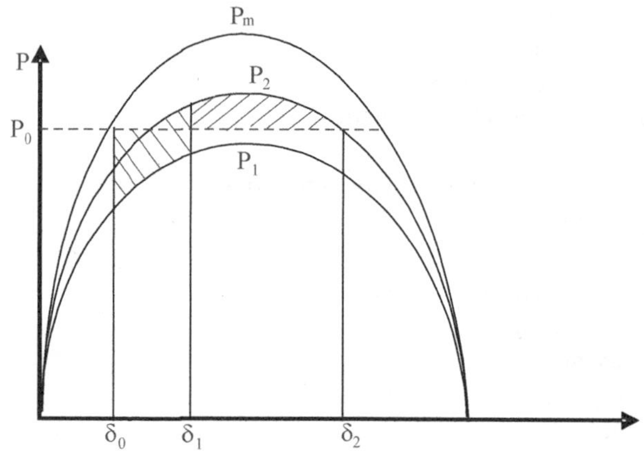

Fig.12.2.1. *Power-angle curves*

$P_0 = 1$ pu

$P_1 = 0.4$ pu

$P_2 = 1.3$ pu

$P_m = 1.8$ pu

$\delta_0 = \sin^{-1}(P_0/P_m) = 0.589$ radian

$\delta_2 = 3.1416 - \sin^{-1}(P_0/P_2) = 2.552$ radian

$\cos\delta_1 = (P_0(\delta_0 - \delta_2) + P_1 * \cos(\delta_0) - P_2 * \cos(\delta_2))/(P_1 - P_2) = 0.611$

$\delta_1 = 52.328°$ **Answer**

Prob.12.2.2 Equal area criterion of stability-Example 2

The transfer reactances between a generator and an infinite bus bar operating at 132 kV , during and after a fault on the interconnector are:

Before the occurrence of the fault =140 Ω / phase

During the fault = 385 Ω per phase

After clearance of the fault = 175 Ω / phase

If the fault is cleared after the generator rotor has advanced 80 electrical degrees from its steady position before the fault , determine the maximum load that could be carried without the fault causing instability.

Solution.

Use the Power-angle Curve of Fig. 12.2.1

Base MVA = 100

V_b, base voltage = 132 kV

I_b base current $= \dfrac{100}{\sqrt{3} \times 132}$

= 0.437 kA

Z_b, = Base kV*Base kV/Base MVA = 174.24 Ω

X_b, reactance before = 0.803 pu

X_d, reactance during = 2.209 pu

X_a, reactance after = 1.004 pu

P_0, Unknown

$P_1 = 1/X_d = 0.452$ pu

$P_2 = 1/X_a = 0.995$ pu

$P_m = 1/X_b = 1.244$ pu

$\delta_1 = 80°$

$\cos\delta_1 = 0.173$

$\cos\delta_1 = (P_0(\delta_0 - \delta_2) + P_1 * \cos(\delta_0) - P_2 * \cos(\delta_2))/(P_1 - P_2)$

$$= P_0 \sin^{-1}\left(\frac{P_0}{P_M}\right) - \pi \sin^{-1}\left(\frac{P_o}{P_2}\right) + P_1 \cos\left(\sin^{-1}\left(\frac{P_0}{P_M}\right)\right) + P_2 \cos\left(\sin^{-1}\left(\frac{P_0}{P_2}\right)\right)$$

In this equation, P_o is the only unknown. It is obtained by trial & error. Selecting a value $P_o = 0.524$, we note that the R.H.S = 0.173

Therefore $P_0 = 0.524$ pu or 52.4 MW is the solution.

Prob.12.2.3 Inertia constant of two generators operating in parallel

A generator A is rated at 50 Hz, 60 MW, 75 MVA, 1500 rpm, and has an inertia constant H = 7.5 MJ/MVA. The corresponding data for another generator B is 50 Hz, 120 MW, 133.3 MVA, 3000RPM, 4.5 MJ/MVA.

(i) If these two generators operate in parallel in a power station, calculate H for the equivalent generator on a base of 100 MVA.

(ii) If the power station is connected to another power station which has two of each type of generator, calculate H for the equivalent generator connected to an infinite bus bar.

Solution. (i)

MVA_{base} $S_{base} = 100$

MVA_a $S_a = 75$

MVA_b $S_b = 133.3$

$H_a = 7.5$ MJ/MVA

$H_b = 4.5$ MJ/MVA

Equivalent H, $He = (H_a.S_a + H_b.S_b)/S_{base} = 11.623$ MJ/MVA

(ii)

H_e1 for the first power station = 11.623 MJ/MVA

H_e2 for the second power station = 2 × He1=23.247 MJ/MVA

H for the equivalent generator connected to an infinite bus bar = $\dfrac{H_{el} + H_{e2}}{H_{el}.H_{e2}}$ = 7.749 MJ/MVA

Answer

Prob.12.2.4 To Determine the Additional Sudden load of an Importing area without loss of Transient Stability.

An importing power area has a total demand of 25 MW from an infinite bus bar via an interconnector. The steady-state power limit is 80 MW. Estimate using the equal-area criterion the additional area load that could be suddenly switched on without the system losing stability.

Solution.

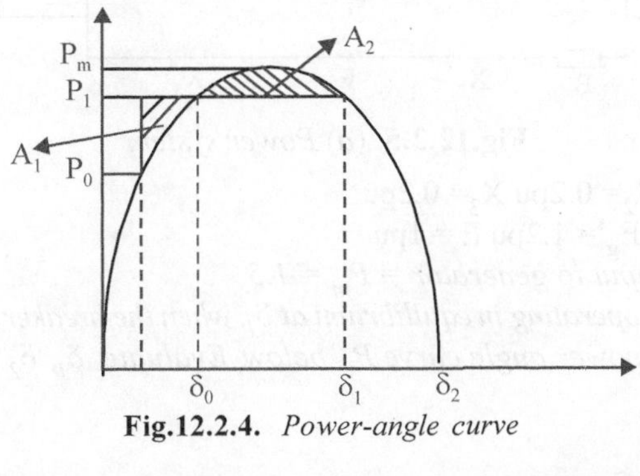

Fig.12.2.4. *Power-angle curve*

$$\text{Area, } A_1 = (\delta_1 - \delta_0)P_1 - \int_{\delta_1}^{\delta_1} P_m \sin\delta.\delta s$$

$$\text{Area, } A_2 = \int_{\delta_1}^{\delta_2} P_m \sin\delta.d\delta - (\delta_2 - \square\delta_1)P_1$$

Equating the two areas we get

$P_1 = P_m(\cos\delta_0 - \cos d_2)/(\delta_2 - \delta_0)$

or, $\sin\delta_1 = (\cos\delta_0 - \cos\delta_2)/(\delta_2 - \delta_0)$ because $P_1 = P_m \sin\delta_1$.

$P_0 = 25$ MW

$P_m = 80$ MW

$$\delta_0 = \sin^{-1}\left(\frac{P_0}{P_M}\right) = 0.317 \text{ rad.}$$

$\cos\delta_0 = 0.949$

$\delta_2 = \pi - \delta_1$

$\cos\delta_2 = \cos\delta_1$

Therefore $\text{Sin}\delta_1 = (\cos\delta_0 - \cos\delta_2)/(\delta_2 - \delta_0)$

$$= \frac{\cos\delta_0 - \cos\delta_1}{2.827 \times \delta_1}$$

$$= \frac{0.999 + \cos\delta_1}{2.827 - \delta_1}$$

We will solve this equation by trial & error.

Try $\delta_1 =$		LHS =		RHS	
	0.5		0.479		= 0.785
	0.6		0.564		0.796
	0.7		0.644		0.805
	0.8		0.717		0.812
	0.9		0.783		0.815
	0.95		0.813		0.815
	0.954		0.815		0.815

$\delta_1 = 0.954$ is the solution. Therefore,

$P_1 = P_m \sin\delta_1 = 65.25886$ MW

Prob. 1.2.5. To use the equal area criterion when the circuit breaker at one end of one of two parallel lines opens suddenly.

Consider the system shown in Fig.12.5 (a).

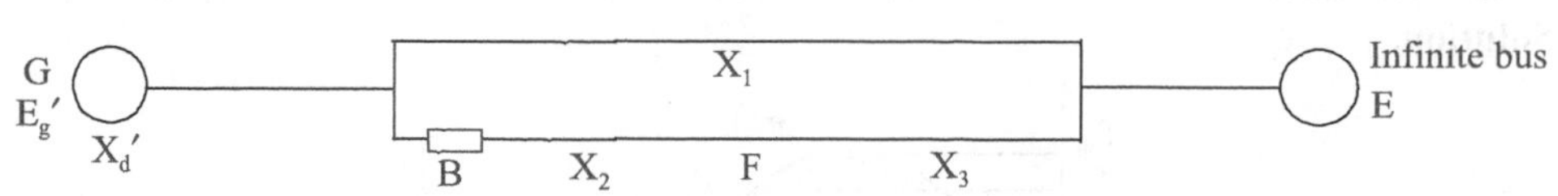

Fig.12.2.5. *(a) Power system*

$X_1 = 0.4$pu $X_2 = 0.2$pu $X_3 = 0.2$pu

$X_d' = 0.2$pu $E_g' = 1.2$pu $E = 1$pu

Mechanical power input to generator $= P_m = 1.5$

Suppose the system is operating in equilibrium at δ_1*, when the breaker B inadvertently, switching the operating point to the power angle curve* P_e' *below. Evaluate* δ_1, δ_2 *and* δ_4.

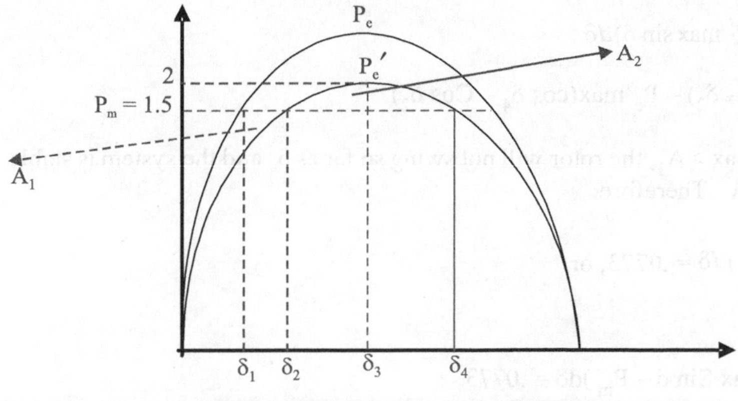

Fig.12.2.5. (*b*) *Power-angle curves*

Solution. At t = 0-, total reactance between the two sources is $X_d' + X$ where

$$X = \frac{X_1(X_2+X_3)}{X_1+X_2+X_3} = 0.2\,pu$$

P_e, max $= \dfrac{E_g' E}{E_d' + X} = 3$pu

$P_e' = P_m$ when $\delta = \delta_1$

i.e., P_e, max $\sin \delta_1 = Pm$, or

$\delta_1 = \sin^{-1}(P_m/P_e, \max) = 29.999°$ **Answer**

At t = 0, B opens, and we have

P_e', max $= E_g' E/(X_d' + X_1) = 2$

$P_m = P_e' \max \sin \delta_2$. Therefore,

$\delta_2 = \sin^{-1}(P_m/P_e' \max) = 48.590°$. **Answer**

$\delta_4 = 131.409°$ (= 180° − δ_2) **Answer**

Prob.12.2.6 To predict the stability /instability of the system of Prob.12.5 when one of two parallel lines opens at one end

Consider Prob.12.5 and predict whether the system will be stable or unstable. If the system is stable, evaluate δ_3.

Solution: $\delta_1 = 30° = 0.523$ rad $\delta_2 = 48.6° = 0.848$ rad $\delta_4 = 131.4° = 2.293$ rad

In the interval δ_1 to $\delta 2$, $P_m > P_e$, and the rotor is accelerating

$$\text{Area } A_1 = \int_{\delta_1}^{\delta_2} (P_m - P_e')\,d\delta$$

$$= \int_{\delta_1}^{\delta_2} (P_m - P_e' \max \text{Sin}\delta)\,d\delta$$

$= P_m(\delta_2 - \delta_1) + P_e' \max (\cos \delta_2 - \cos \delta_1)$

$= 0.0775$

The system will be stable when there is enough area A_2 in the interval $\delta_2 < \delta < \delta_4$ to offset the acceleration.

$$A_2 \max = \int_{\delta_2}^{\delta_4} (-P_m + P_e')\,d\delta$$

$$\int_{\delta_2}^{\delta_4}(-P_m + P_e' \max \sin \delta) d\delta$$

$= - P_m(\delta_4 - \delta_2) - P_e'\max(\cos \delta_4 - \cos \delta_2)$
$= 0.477$

Since $A_2\max > A_1$, the rotor will not swing so far as δ_4 and the system is stable. The rotor will stop when $A_2 = A_1$. Therefore,

$$\int_{\delta_2}^{\delta_3}(P_e' - P_m) d\delta = .0773, \text{ or}$$

$$\int_{\delta_2}^{\delta_3} (P_e'\max \sin \delta - P_m) d\delta = .0773$$

which gives

$-P_e'\max (\cos \delta_3 - \cos \delta_2) \; -P_m(\delta_3 - \delta_2) = .0773$

or,

$1.5\,\delta_3 + 2 \cos \delta = 2.518$

Use iterative method to solve this non-linear equation.

	$\delta_3 =$ 1	$\delta_3 =$ 1.1	$\delta_3 =$ 1.2	$\delta_3 =$ 1.2 1	$\delta_3 =$ 1.22	$\delta_3 =$ 1.218
$1.5\,\delta_3 + 2 \cos \delta_3 =$	2.580	2.557	2.524	2.521	2.5127	2.518

Therefore solution is :

$\delta_3 = 1.218$ rad
$= 69.786°$.

Answer

Prob.12.2.7 To predict the stability /instability of the system of Prob.12.2.5 when fault occurs at the midpoint of one of two parallel lines.

Suppose a balanced three-phase fault occurs at F in the power system of Prob.12.2.5.
(a) If the fault is not removed, will the generator be stable ?
(b) If the fault is cleared by opening the breaker B at $\delta_c = 60°$; will the system be stable? If so, calculate δ_3.

Solution: (a) With the fault, the system becomes as shown in Fig.12.7 (a).

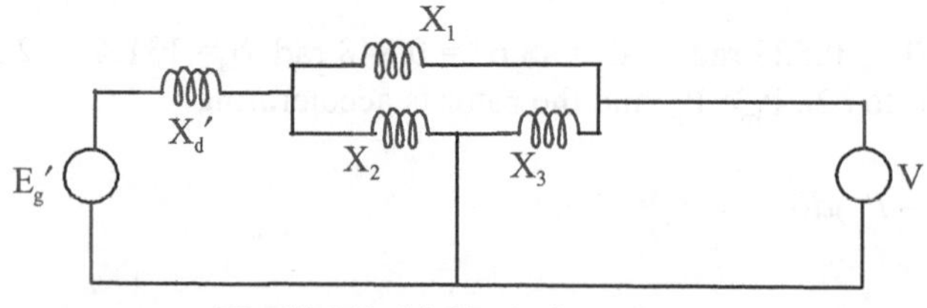

Fig.12.2.7. *(a) Equivalent circuit*

This can be reduced to Fig.12.7(b) :

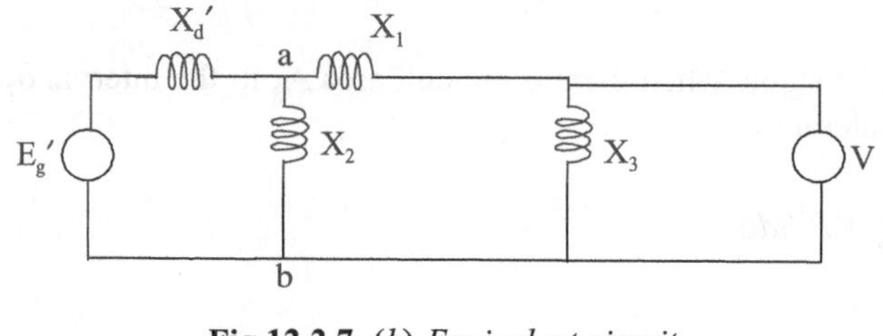

Fig.12.2.7. *(b) Equivalent circuit*

$E_g' = 1.2$ pu ; $X_d' = 0.2$ pu ; $X_1 = 0.4$ pu
$V = 1$ pu ; $X_2 = 0.2$ pu ; $X_3 = 0.2$ pu

The Thevenin equivalent of the circuit across a-b is shown in Fig.12.7(c):

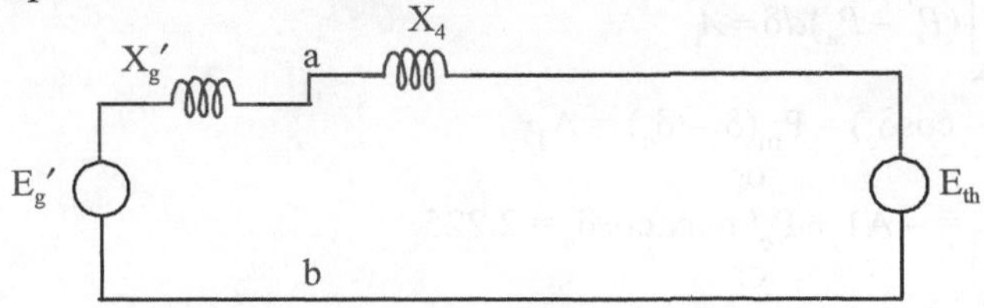

Fig. 12.2.7. (*c*) *Equivalent circuit*

$$X_4 = \frac{X_1 + X_2}{X_1 + X_2} = 0.133 \text{ pu}$$

$$E_{th} = \frac{VX_2}{X_1 + X_2} = 0.333 \text{ pu}$$

Under faulted condition, $P_e'' = P_e''\max \sin\delta$

$$P_e''\max = \frac{E_g' E_{th}}{X_d' + X_4} = 1.2 \text{ pu}$$

Stability is impossible, since $P_e''\max$ (=1.2) is less than P_m (=1.5), preventing even steady-state stability.

(b)

$P_e\max$ = Pre-fault = 3
$P_e'\max$ = Post-fault = 2
$P_e''\max$ = Faulted = 1.2

If the fault is cleared at $\delta_c = 60° = 1.0472$ rad.
$\delta_1 = 30° = 0.523$ rad. $P_m = 1.5$ pu $\delta_4 = 131.4°.=2.293$ rad.

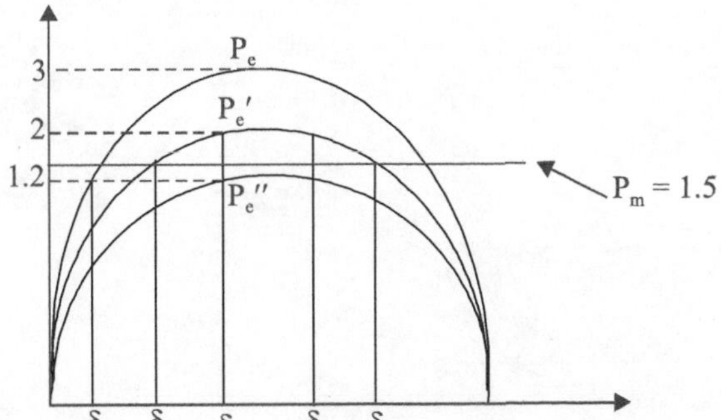

Fig. 12.2.7 (*d*). *Power angle curves*

$$A_1 = \int_{\delta_1}^{\delta_c} (P_m - P_e'')\, d\delta = P_m(\delta_c - \delta_1) + P_e''\max(\cos\delta_c - \cos\delta_1) = 0.346 \text{ pu}$$

Maximum available decelerating area is :

$$A_2\max = \int_{\delta c}^{\delta_4} (P_e'\max \sin\delta - P_m)\, d\delta = - P_e'\max(\cos\delta_4 - \cos d_c) - P_m(\delta_c - \delta_c) = 0.453 \text{ pu}$$

Since $A_2max > A_1$, system is stable.

To solve for δ_3,

$$\int_{\delta c}^{\delta_3}(P_e' - P_m)d\delta = A_1$$

$$-P_e'max(\cos\delta_3 - \cos\delta_c) - P_m(\delta_3 - \delta_c) = A_1$$

or

$$1.5\ \delta_3 + 2*\cos\delta_3 = -A1 + P_e'\ max.\cos\delta_c = 2.225$$

or

$$1.5\ \delta_3 + 2*\cos\delta_3 = 2.225$$

This non-linear eq. is solved below :

$\delta_3 =$	1	1.5	1.848 radian
$1.5\delta_3 + 2*\text{Cos}\delta_3 =$	2.580	2.391	2.224

The solution is :

$\delta_3 = 1.848$ radian **Answer**

CHAPTER 13

OVERVOLTAGES

13.1 INTRODUCTION

Look up the following Power Related Software at http://www.mtu.edu/atp :

ATP, Alternative Transients Program

The transient overvoltages that occur on a power system are either of external origin (*e.g.*, a lightning discharge) or are generated internally by switching operations. In general, the transients in transmission systems are cause by any sudden change in the operating condition or configuration of the systems. Lightning is always a potential hazard to power system equipment, but switching operations can also caüse equipment damage. At voltages up to about 230 kV the insulation level of the lines & equipment is dictated by the need to protect against lightning. On systems where voltages are above 230 kV but less than 700 kV switching operations as well as lightning are potentially damaging to the insulation. At voltages above 700 kV switching surges are the main determinant of the level of insulation.

13.1.1 Travelling Waves

The parameter of the transmission lines, generator & transformer windings are distributed. A distinguishing feature of the circuit with distributed parameters is its ability to support travelling waves of current & voltage. The velocity of propagation of current & voltage waves along a lossless line is

$$v = 1/\sqrt{(LC)} = 1/\sqrt{(\varepsilon_{oo})} = 3\times 10^8 \text{ m/sec} = \text{velocity of light for an overhead line.}$$

In some practical cases, the ambient medium is not free space or air. For example, in a cable the permittivity of the dielectric is $\varepsilon_o\varepsilon_r$ where ε_r may be 3 to 5 or even more. Thus, the velocity of wave propagation in cable is reduced by a factor $1/\sqrt{\varepsilon_r}$. It may be less than half the velocity for an air insulated line. Similarly, where the conductor is buried in steel, as are the conductors of most rotating machines, the permeability $\mu(=\mu_o\mu_r)$ may greatly exceed μ_o.

The ratio of the amplitudes of voltage and current waves is called the characteristic impedance Z_o of the line.

$$Z_o = \sqrt{(L/C)}$$

For a typical overhead transmission line, Z_o is about 400 Ω. For an underground cable it is in the range of 30-80 ohms, because the closer spacing of the conductors makes C larger & L smaller.

13.1.2 Reflection & Refraction of Travelling Waves

There is a strict proportionality between the voltage and the current waves in a transmission line. The proportionality factor is the characteristic impedance Z_o of the line. When a wave arrives at a discontinuity in a line where the characteristic impedance of the line changes, some adjustment must occur if this proportionality is not to be violated. This adjustment takes place in the form of the initiation of two new wave pairs. The reflected voltage wave and its companion current wave travel back down the line superimposed on the incident wave. The refracted wave penetrates beyond the discontinuity. The amplitudes of the reflected & refracted waves are such that the voltage to current proportionalities are preserved for each, as demanded by the characteristic impedances of the lines on which they are travelling.

Consider the junction between the lines of characteristic impedances Z_a and Z_b. This may be a junction between an overhead line and a cable. If a step function surge of amplitude V_1 approaches the junction along the overhead line then the incident (V_1), reflected (V_2) and refracted (V_3) voltage waves are given by

$$V_2 = \frac{(Z_b - Z_a)V_1}{Z_a + Z_b}$$

$$V_3 = \frac{2Z_b V_1}{Z_a + Z_b}$$

The quantity $\frac{Z_b - Z_a}{Z_b + Z_a}$ is called the reflection coefficient while the quantity $\frac{2Zb}{Za + Zb}$ is called the refraction coefficient

The reflected current, $I_2 = \frac{-V_2}{Z_a}$

The refracted current $I_3 = \frac{-V_3}{Z_b}$

- Study the behaviour of travelling waves at
  1. An open-circuited line
  2. A short-circuited line, and
  3. A line terminated in a resistive load, R

13.1.3 Bewley Lattice Diagram

The Bewley lattice diagram is a space-time diagram with space measured horizontally and measured time vertically. This diagram show at a glance the position and direction of motion of every incident, reflected and transmitted wave on the line at every instant of time. By its means the difficulty of keeping track of the multiplicity of successive reflections is considerably simplified.

- Illustrate the use of the Bewley lattice diagram by using the following examples :
  1. An open-circuited lossless line energized with step input voltage V_1
  2. A short-circuited lossless line energized with step input voltage V_1
  3. A lossless line terminated in a resistive load, R
  4. A long line connected to a cable of finite length with its remote end open-circuited.

- What do you understand by a distortion-less line ?
- What is the attenuation constant of a distortion-less line? What is the power (and energy) associated with a distortion-less line ?
- Show that for a transmission line the attenuation constant α is

$$\alpha = \frac{R}{2Z} + \frac{GZ}{2}$$

where, R = resistance per unit length
Z = Characteristic impedance
G = leakage conductance

13.1.4 Transient Recovery Voltage

The circuit in which the breaker is located plays a very significant part in the interruption process, since if the rate of rise of voltage across the contacts immediately after a current zero, particularly in low-resistance interruption, exceeds the rate of rise of the dielectric strength of the contact space, the arc will be reformed and continue for a further period. This transient voltage is normally oscillatory due to the presence of inductance and stray capacitance; it can however, be damped by circuit resistance or by resistors installed for the purpose. Another factor is current chopping, *i.e.*, the sudden interruption of the current before a natural current zero; this is liable to occur with air-blast breakers interrupting low or medium currents or with vacuum breakers. Example 13.2.5 shows a typical transient voltage calculation.

OBJECTIVE TYPE QUESTIONS

1. Which one of the following statements is true ?
   In a protection scheme for a communication line,
   (*a*) The lightning arrester prevents any accumulation of static charge , and the drainage coil relieves the system of any excess voltage.
   (*b*) The lightning arrester relieves the system of any excess voltage , and the drainage coil prevents any accumulation of static charge.
   (*c*) The lightning arrester prevents any accumulation of static charge on the communication circuit.
   Ans. (*b*)
2. The correct meaning of 50% dry flash-over test is:
   A positive 1/50 μsec.wave is applied 20 times to the insulator. Then a negative 1/50 microsecond wave is applied to the same insulator 20 times.
   (*a*) The amplitude of the wave is such that the insulator flashes over in each case, but is not damage.
   (*b*) The amplitude of the wave is such that the insulator is damaged in 50 % of the cases.
   (*c*) The amplitude of the wave is such that 50% of the positive impulses cause flashover and 50 % of the negative impulses cause flashover, but the insulator is not damaged.
   Ans. (*c*)

PROBLEMS WITH SOLUTIONS

Prob.13.2.1 Surges in two transmission lines connected by an underground cable

The ends of two long overhead lines A and B are connected by a cable C 1.5 km long. The lines have capacitance 10 pF/m and inductance 1.6 × 10E-07 H/m and the cable has capacitance 89 pF/ m and inductance 5 × 10E-08. A rectangular voltage wave of magnitude 10 kV and of long duration, travels along line A towards the cable. Find the magnitude of the second voltage step occurring at the junction of the cable and the line B. What will be the voltage at the junction of line A and the cable, 20 μsec. after the initial surge reaches this point?

Solution.

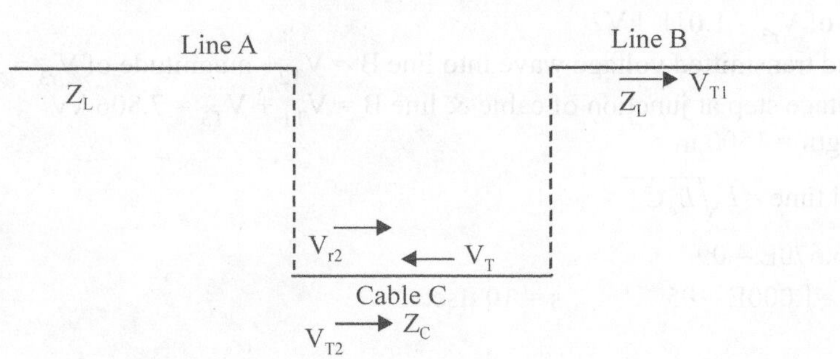

Fig.13.2.1. (*a*) *System and Travelling waves*

L_l line inductance = 0.0000016 H/m
C_l line capacitance = 0.00000000001 F/m
L_C cable inductance = 0.0000005 H/m
C_c cable capacitance = 0.000000000089 F/m

Z_l line surge impedance = $\sqrt{\dfrac{L_1}{C_1}}$ = 400 Ω

Z_c cable surge impedance = $\sqrt{\dfrac{L_C}{C_C}}$ =74.953 Ω

K_{r1} Reflection coefficient = $\dfrac{Z_c - Z_1}{Z_c + Z_p}$ = – 0.684
V_i Incident voltage = 10.0 kV
V_T Transmitted voltage into cable = $(1+ K_{r1})*V_i$= 3.156 kV

V_{T1} Transmitted voltage into line B = $\dfrac{2Z_1 V_T}{Z_c + Z_l}$ = 5.316 kV

V_r2 Reflected voltage from line B = $\dfrac{(Z_1 - Z_c)}{Z_l + Z_c} V_T$ = 2.160 kV

V_{T2} Reflection of V_{r2} at the junction of cable and line A = $\dfrac{Z_e - Z_c}{Z_e + Z_c} V_{r2}$ =1.478 kV

Lattice diagram :

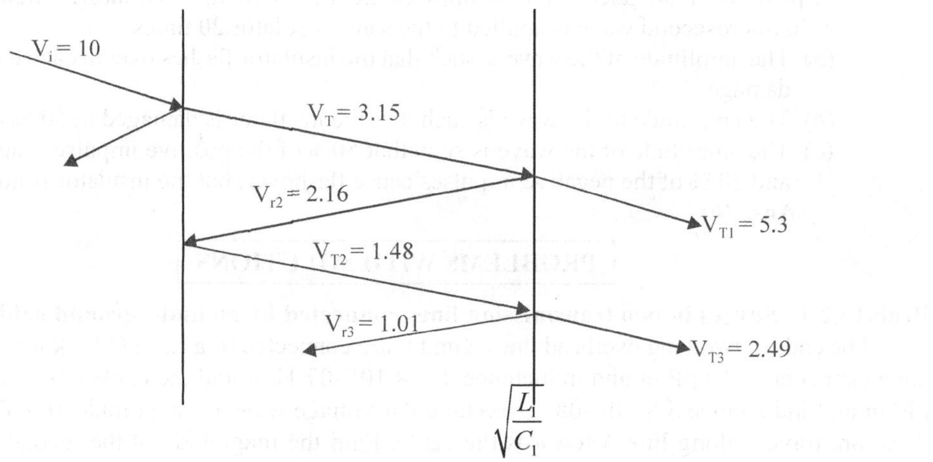

Fig.13.2.1. (*b*) *Travelling waves*

V_{r3}, Second reflection from line B = $K_{r1} \times V_{T2}$ = – 1.011 kV
Magnitude of V_{r3} = 1.011 kV
V_{T3}, Second transmitted voltage wave into line B = V_{T2}+ magnitude of V_{r3} = 2.489 kV
Second voltage step at junction of cable & line B = $V_{T1} + V_{T3}$ = 7.806 kV **Answer**
l, cable length = 1500 m

cable travel time = $l\sqrt{L_c C_c}$

$\sqrt{L_c C_c}$ = 6.670E – 09
travel time =1.000E – 05 s = 10 μsec.

Voltage at the junction of line A and cable, 20 microseconds after the initial surge reaches that point = $V_T + V_{r2} + V_{T2}$ = 6.794 kV **Answer**

Prob.13.2.2

Surges in a transmission line connected to an underground cable through a resistor

An overhead line of surge impedance 400 Ω is connected to a cable of surge impedance 40 Ω through a series resistor as shown in Fig.13.2.2 (a). Determine the magnitude of the resistor such that it absorbs maximum energy from a surge originating on the overhead line and travelling into the cable. Calculate

(a) the voltage and current transients reflected back into the line.

(b) those transmitted into the cable in terms of the incident surge voltage of 1000 kV, and

(c) the energies reflected back into the line and absorbed by the resistor.

Solution.

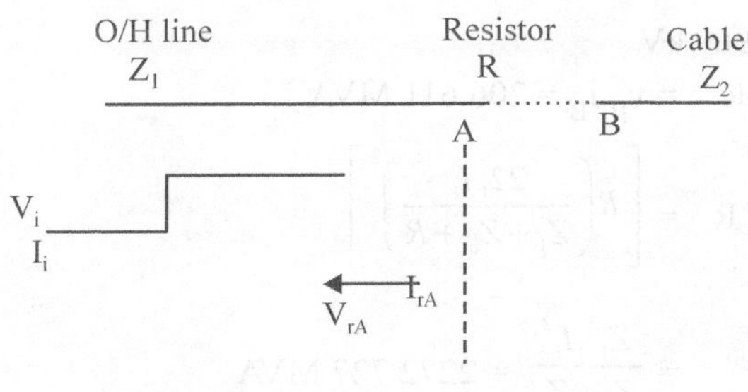

Fig.13.2.2. (*a*) *Line and cable diagram*

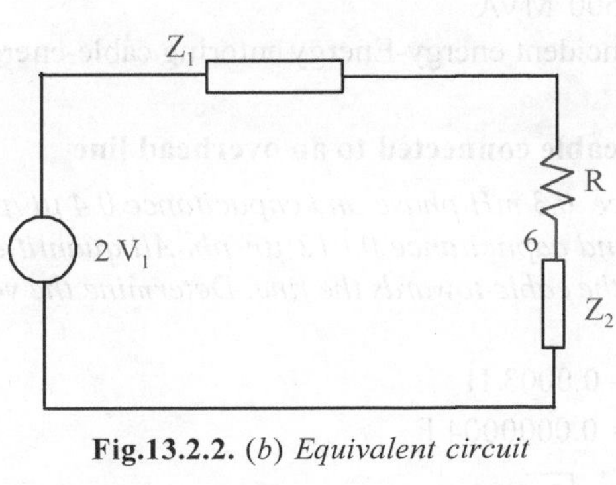

Fig.13.2.2. (*b*) *Equivalent circuit*

$$V_B = \frac{2V_1Z_2}{Z_1+Z_2+R}$$

$$V_{rA} = \frac{Z_2+R-Z_1}{Z_1+Z_2+R}V_i)$$

$$I_B = \frac{2V_i}{Z_1+Z_2+R}\ \frac{2Z_1I_i}{Z_1+Z_2+R}$$

$$I_{rA} = \frac{I_i(Z_1+Z_2+R)}{Z_1+Z_2+R}$$

Power absorbed by R = $R.I_i^2$

$$= R\left(\frac{2Z_1I_i}{Z_1+Z_2+R}\right)^2$$

It is maximum when $\frac{d}{dR}\left[R\left(\frac{R}{Z_1+2_a+R}\right)^2\right] = 0$, if I_i is given and constant.

This gives the condition that

$R = Z_1 + Z_2 = 440\ \Omega$

$V_i = 1000$ kV

$I_B = \frac{2V_i}{Z_1+Z_2+R} = 2.272$ kA **Answer**

$I_i = V_i/Z_1 = 2.5$ kA

$I_{rA} = -0.227$ kA **Answer**

$V_B = 90.909$ kV **Answer**

$V_{rA} = V_B - V_i = 90.909$ kV **Answer**

Energy entering cable $= v_B.i_B = 206.611$ MVA

Energy absorbed by R $= \left[R\left(\frac{22_1 I_i}{Z_1+Z_a+R}\right)^2\right]$

$= \frac{Z_1^2.I_i^2}{Z_1+Z_2} = 2272.727$ MVA **Answer**

Energy incident = 2500 MVA

Energy reflected = Incident energy-Energy entering cable-energy absorbed = 20.661 MVA **Answer**

Prob.13.2.3 Surges in a cable connected to an overhead line

A cable of inductance 0.3 mH/phase and capacitance 0.4 μF/phase is connected to a line of inductance 1.5 mH /ph and capacitance 0.012 μF/ph. All quantities are per km. A surge of 1pu magnitude travels along the cable towards the line. Determine the voltage set up at the junction of the line and the cable.

Solution. $L_c = 0.0003$ H

$C_c = 0.0000004$ F

$Z_c = \sqrt{\frac{L_C}{C_C}} = 27.386\ \Omega$

$L_l = 0.0015$ H

$C_l = 0.000000012$ F

$Z_l = \sqrt{\frac{L_l}{C_l}} = 353.553\Omega$

K Reflection coefficient = 0.856

Voltage at junction = 1+ k = 1.856 pu **Answer**

Prob.13.2.4 Surges in two transmission lines connected by an inductor

A rectangular surge of 2 microsecond duration and magnitude 2 pu travels along a line of surge impedance 350 ohms. The latter is connected to another line of equal impedance through an inductor of 800 micro-Henries. Calculate the value of the surge transmitted to the second line.

Solution.

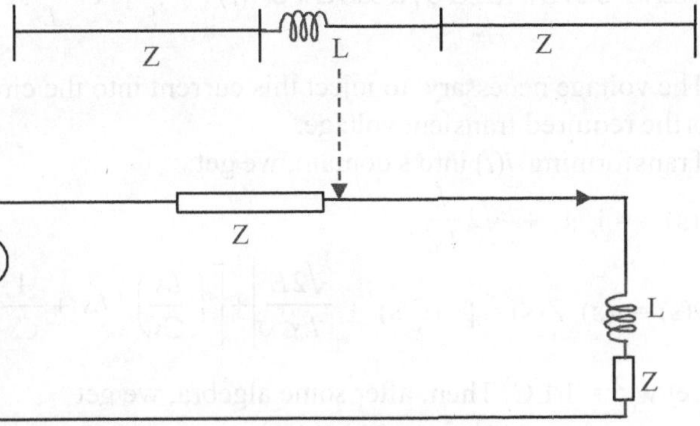

Fig.13.2.4. *System diagram and Equivalent Circuit*

v_i = 2 pu
Z = 350 Ω
L = 0.0008 H
t = 0.000002 s

$$I = \left(\frac{2Vi}{2Z}(1-e^{-2\frac{Zt}{L}} \right)$$

V in the second line = I × Z

= $v_i(1-\exp(-2Z\,T/L))$ = 1.652 pu **Answer**

Prob. 13.2.5 Transient Recovery Voltage

A 1-phase, 50 Hz generator with one terminal earthed is excited to an internal emf of 10 kV and is connected to a system through a circuit breaker. The generator inductance is 10 mH and the effective stray capacitance is represented by a 4 μF capacitor across the circuit breaker. A short- circuit occurs beyond the circuit breaker. Develop an expression for the voltage across the contacts immediately after the current interruption if this occurs at an instant corresponding to 10° prior to a natural current zero.

Solution.

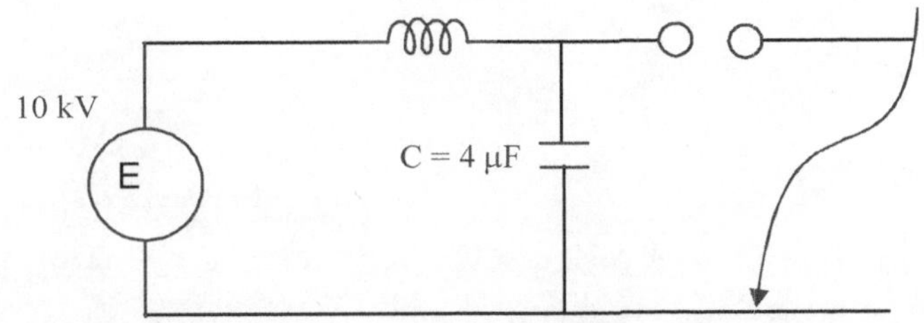

Fig. 13.2.5. Circuit diagram

E = 10 kV = 10000 V
ω = 314 rad/s
L = 0.01 H
C = 0.000004 F

Short circuit current = E/ωL = 3184.713 A (r.m.s)

Rate of change at 10° (assumed same as at zero) = $\frac{\sqrt{2}E}{L_s}$ = 1414214 A/s value of current at 10° = i_c =2 × 3180 × 0.123 = 782.28 A

To bring the breaker current to zero, a current of $i(t) = -i_c + \sqrt{2}\frac{E}{L}.t$ is imagined to be injected across the contacts.

The voltage necessary to inject this current into the circuit impedance is the required transient voltage.

Transforming $i(t)$ into s domain, we get:

$$i(s) = -i_c/s + \sqrt{2}\frac{E}{Ls^2}$$

$$v(s) = i(s).\ Z(s) = \left[-(i_c/s) + \frac{\sqrt{2}E}{Ls^2}\right] + \left[\left(\frac{Ls}{Cs}\right) / Ls + \frac{1}{Cs})\right]$$

Let $w_0^2 = 1/LC$. Then, after some algebra, we get

$$V_{(s)} = i_c\sqrt{\frac{L}{C}}.\frac{\omega_0}{s^2+\omega_0^2} + \frac{2E\omega_0^2}{L.s(s^2+\omega_0^2)}$$

In time domain, the above equation becomes

$$V(t) = i_c\sqrt{\frac{L}{C}}\sin\omega_0.t + \sqrt{2}E(1-\cos\omega_0 t)$$

$$\omega_0 = \sqrt{\frac{1}{LC}} = 5000$$

$$\sqrt{LC} = 50\ \Omega$$

$$i_c\sqrt{\frac{L}{C}} = 39114$$

$$\sqrt{2}E = 14142.14$$

v(t) = – 39114 sin 5000t + 14142 (1– cos 5000 t)

CHAPTER 14

AUTOMATIC GENERATION CONTROL

13.1 INTRODUCTION

Look up the websites:

- http://dharma.ece.uiuc.edu/pwrworld.html for the software: PowerWorld dealing with Power Systems operation & control.
- http://powerlearn.ee.iastate.edu for the software : Animation of mechanical-hydraulic speed governor for power generator

14.1.1 Integrated operation of power systems

- Need for integrated operation of power systems
- Review of load sharing process between two generators
- The power/ frequency characteristic of interconnected system.
- Interconnection of systems through lines of relatively smaller capacities
- Use of phase shifting transformers
- Exchange of power between systems
- Frequency & tie line power control
- The concept of power pools

OBJECTIVE TYPE QUESTIONS

1. Two power systems A and B are operating in parallel. If system A increases its generation to deliver 100 MW to system B, what will be the effect if system B does not reduce its generation simultaneously ?

(*a*) frequency will increase

(*b*) frequency of both A and B will decrease

(*c*) frequency will remain unchanged.

Ans. (*a*)

2. Governors for controlling the speed of power generation units provide

(*a*) A flat speed -load characteristics

(*b*) An increase of speed with increasing load

(*c*) A decrease of speed with increasing load

Ans. (*c*)

3. When two identical a.c generating units are operated in parallel on governor control, and one machine has a 5% governor drop and the other a 10 % droop , the machine with the greater governor droop will

(*a*) Tend to take the greater portion of load changes

(*b*) Share load equally with the other machine

(*c*) Tend to take the lesser portion of load changes.

Ans. (*c*)

4. On load-frequency control installations, error signals are developed proportional to the frequency error. If frequency declines , the error signal will act to

(*a*) Increase prime move input to the generators
(*b*) Reduce prime mover input to the generators
(*c*) Increase generator voltages
Ans. (*a*)

5. When inter-connected power systems operate with tie-line bias, they will respond to
(*a*) Frequency changes only
(*b*) Both frequency & tie -line load changes
(*c*) Tie-line load changes only
Ans. (*b*)

6. Power system A is operating in parallel with other power systems and has a frequency bias of 50 MW per 0.1 Hz. For a loss of generation in one of the interconnected systems which results in a frequency drop of 0.03 Hz , power system A should
(*a*) Provide 30 MW of support
(*b*) Not be affected
(*c*) Provide 15 MW of support
Ans. (*c*)

7. Area control error is given by ACE = $(T_o - T_1)$ - 10 .B. $(f_0 - f_1)$, where
ACE- area control error
T_o -scheduled net interchange
T_1- Actual net interchange
f_o- desired frequency, Hz
f_1-actual frequency, Hz
B- Area bias, MW/one-tenth Hz
If scheduled net interchange T_o = 200 MW into the system, actual net interchange T_1 =180 MW into the system, f_o = 50 Hz, f_1= 49.9 Hz, B = 150 MW/0.1 Hz, then the ACE is
(a) + 150 MW
(b) zero
(c) +170 MW
(d) – 130 MW
Ans. (*d*)

8. If frequency bias is set too low, an interconnected power system will respond to troubles in adjacent systems with
(*a*) More than its share of bias response
(*b*) Correct bias response
(*c*) Less than its share of bias response
Ans. ( )

9. If a power system operates at an average frequency of less than 50 Hz,
(*a*) No time error will accumulate
(*b*) There will be a slow accumulated time error
(*c*) There will be a fast accumulated time error.
Ans. (*b*)

10. When a power system operator observes an accumulated time error, he should correct it by
(*a*) Increasing generation on his system
(*b*) Reducing generation on his system
(*c*) Co-ordinating time error with other interconnected systems
Ans. (*c*)

PROBLEMS WITH SOLUTIONS

Prob.14.2.1 Governor Characteristics & load distribution

Fig. 14.2.1 shows two 500 kW alternators running in parallel. The governors on the prime movers are so set that the frequency drops from 51 Hz to 49 Hz from no load to full load in one, while in the other from 51 to 49.5 Hz. Calculate

(a) the load on each machine when the total load is 750 kW, and

(b) the frequency at this load

Solution.

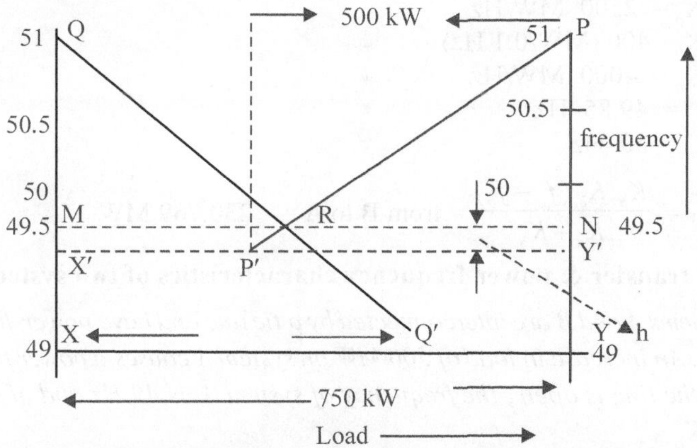

Fig.14.2.1. *Loads on the two alternators*

The intersection point R of the two load lines gives the loads on the two machines.

$$MR = \left(\frac{MQ}{XQ}\right) XQ' = \frac{(1.5-h)\times 500}{2}$$

Triangles NPR and $Y'PP'$ are similar.

$$NR = NY\frac{Y'P'}{Y'P} = (1.5-h)\frac{500}{1.5}$$

$$MR + NR = 750 = (1.5-h)\left(\frac{500}{2}+\frac{500}{1.5}\right)$$

$h = 1.5 - 750/(250 + 333.3) = 0.214$

$MR = (1.5 - h) \times 500/2 = 321.5$ kW **Answer**

$NR = 750 - 321.5 = 428.5$ kW **Answer**

Frequency = 49.5 + 0.214 = 49.714 **Answer**

Prob.14.2.2 Power transfer & power frequency characteristics of two systems

A power system A has a power frequency characteristic such that a drop in frequency of 0.1 Hz is caused by an additional load of 250 MW, and the corresponding load change on adjacent network B for the same frequency deviation is 400 MW. At a certain time A is running at 49.85 Hz and B at 50 Hz. What would be the power transfer if the two systems were to be paralleled via a short a.c. interconnector ?

Solution.

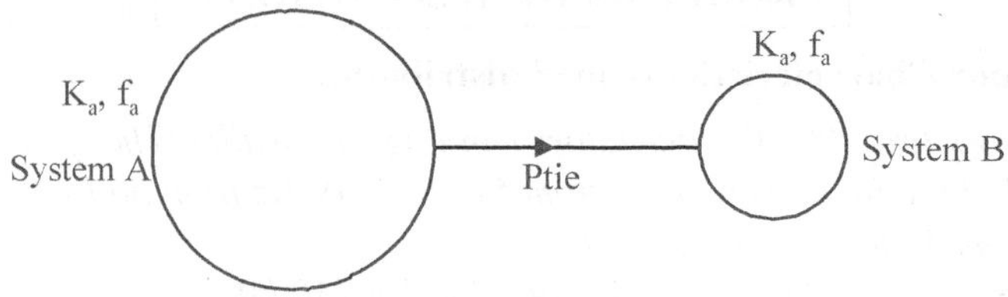

Fig.14.2.2. *Two systems connected by a tie-line*

Load frequency characteristics:

$K_a = 250(\text{MW}/0.1\text{ Hz})$
$K_a = 2500\ \text{MW/Hz}$
$K_b = 400\ (\text{MW}/0.1\text{ Hz})$
$K_b = 4000\ \text{MW/Hz}$
$f_a = 49.85\ \text{Hz}$
$f_b = 50\ \text{Hz}$

P tie, Tie line flow = $\dfrac{K_a . K_b\,(f_a - f_b)}{K_a + K_b}$ = from B to A = – 230.769 MW **Answer**

Prob.14.2.3 Power transfer & power frequency characteristics of two systems-Example 2

Two power systems A and B are interconnected by a tie line and have power-frequency constants K_A and KB, MW/Hz. An increase in load of 500 MW on system A causes a power transfer of 300 MW from B to A. When the line is open , the frequency of system A is 49 Hz and of system B is 50 Hz. Determine K_A and K_B.

Solution.

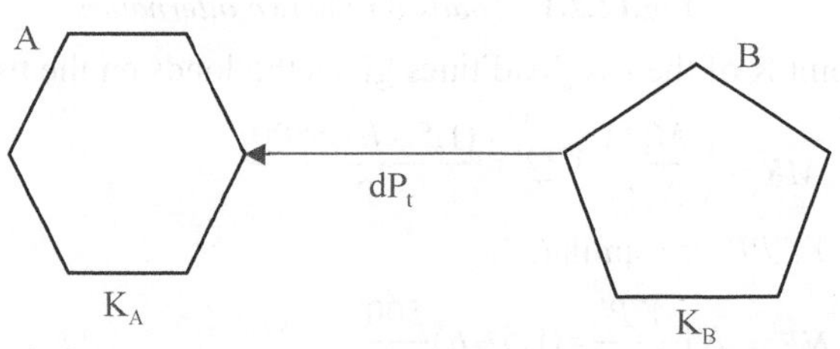

Fig.14.2.3. (*a*) *Two systems connected by a tie-line*

Change in frequency in A due to extra load dP and extra input dP_t from B = – $(dP - dP_t)/K_A$

Drop in frequency in B due to extra load = – dP_t/K_B

From the two changes in frequency, we get

$$dP_t = \frac{K_B . dP}{K_A + K_B} \qquad \ldots (1)$$

Now open the link:

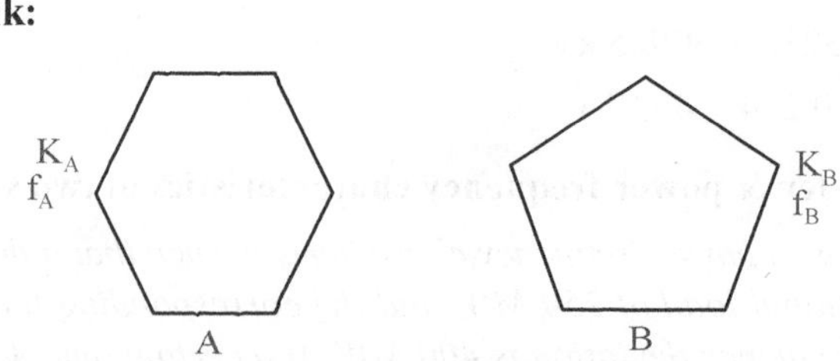

Fig.14.2.3. (*b*) *Tie-line link opened*

B is relieved of dP_t and assumes f_B

A has dP_t more load and assumes f_A.

$f_B = f + dP_t/K_B$ ... Eq.1

$f_A = f - dP_t/K_A$

$$dP_t = \frac{(f_B - f_A)\,K_B K_A}{K_A + K_B} \quad \text{... Eq.2}$$

Eliminating $K_B/(K_A+K_B)$ from eqs.1& 2, we get

$K_A = dP/(f_B - f_A)$ ... Eq.3

$dP = 500$ MW

$f_B = 50$ Hz

$f_A = 49$ Hz

Therefore,

$K_A = 500$ MW/Hz **Answer**

$dP_t = 300$ MW

From Eq.(2) we get

$$K_B = \frac{K_A . dP_t}{dP - dP_t} = 750 \text{ MW/Hz}$$ **Answer**

Prob.14.2.4 Frequency & Tie-line power errors for a load change in a 2-Area system

Two power systems A and B , each have regulation (R) of 0.1 pu (on respective capacity bases) and a stiffness K of 1 pu. The capacity of system A is 150 MW and of B is 100 MW. The two systems are interconnected through a tie line and are initially at 50 Hz. If there is a 10 MW load change in system A, calculate the change in the steady-state values of frequency and power transfer.

Solution: P_A = 150 MW

P_B = 100 MW

K_A = 150 MW/Hz

K_B = 100 MW/Hz

$R_A = R_B$ = 0.1 pu

f = 50 Hz

R_A, Hz/MW = change in frequency-no Load to full Load/full Load capacity = $R_A(\text{pu}) \times f/P_A$ Hz/MW

= 0.033 Hz/MW

= $R_B(\text{pu}) \times f/P_B$ Hz/MW

R_B, Hz / MW = 0.05 Hz/MW

= 10 MW

Steady-state error in frequency = – change in load/$(K_A + (1/R_A)) + (K_B + (1/R_B))$ = – 0.033 Hz **Answer**

Steady-state error in tie-line flow = – change in load$(K_B + (1/R_B))/(K_A+(1/R_A) + K_B + (1/R_B))$ = – 4 MW **Answer**

CHAPTER 15

CONTROL OF VOLTAGE & REACTIVE POWER

15.1 INTRODUCTION

Relation between voltage, power, and reactive power at a node:

Voltage drop between two nodes1 and 2, at voltages V_1 and V_2 respectively, connected by a short transmission line of impedance $R + jX$ is $= (RP_2 + XP_2)/V_2$ where P_2, Q_2 is the real and reactive power at node V_2. For most power networks X >> R and the voltage drop determines Q.

If V_1 is in phase advance of V_2, then the power P flows from node1 to node 2.

If $V_1 > V_2$, then reactive power is transferred from node1 to node 2.

If by varying the excitation of generators at nodes 1 & 2 V_2 is made $> V_1$, then the direction of Q will be reversed from node 2 to node1.

Hence, P can be sent from node 1 to node 2 or from node 2 to node1 by suitably adjusting the amount of steam (or water) admitted to the turbine and Q can be sent in either direction by adjusting the voltage magnitudes. These two operations are approximately independent of each other if X>>R.

Generation & Absorption of Reactive Power

If a scalar voltage difference exists across a reactive link, the reactive power flows towards the node of lower voltage. Or, if there is a deficiency of Q at a point, this has to be supplied from the connecting lines and hence the voltage at that point falls. If there is a surplus of Q generated (lightly loaded cables absorb leading or negative Vars and hence generate positive Vars) then the voltage will rise.

Voltage can be controlled at a node by injecting into the node a Q of the correct sign. Other methods of controlling the voltage are the use of tap changing transformers and voltage boosters.

Synchronous Generators

These can be used to generate or absorb Q. The ability of the generator to supply Q is determined by the short circuit ratio (S.C.R $= \frac{1}{X_s}$). In modern machines SCR is made low for economic reasons and hence the inherent ability of the machine to operate at leading power factors is not large. The Var capacity of the generator can be increased by the use of continuously acting voltage regulators. An overexcited machine generates reactive power. An underexcited machine absorbs (or generates negative or leading) Vars. The generator is the main source of supply to the system of both positive & negative Vars.

Overhead Lines & Transformers

When fully loaded an overhead line absorbs Q (= I^2X per phase). On light loads, the shunt capacitance of long lines may become predominant and the lines become Var generators.

Transformers always absorb Q.

Vars absorbed by a transformer = (VA of load)^2. Xpu/Rated VA

where Xpu is the pu reactance of the transformer.

Underground cables are generators of Q owing to their high capacitance.

Loads

A load of 0.95 pf lagging implies a Q demand of 0.33 kVar per kW of power. In planning a network it is desirable to assess the Q requirements to ascertain whether the generators are able to operate at the required power factors for the extremes of load to be expected.

15.1.2 Injection of Reactive Power -Static & Synchronous Condensers

Three methods of injecting Q at the loads are :

1. Static shunt capacitors
2. Static series capacitors, and
3. Synchronous compensators

Shunt Capacitors & Reactors

Shunt capacitors are used for lagging power factor circuits. Shunt capacitors are disposed along routes to minimize losses and voltage drops. On light loads, when the voltage is high, the capacitor output is large and the voltage tends to rise to excessive levels.

Shunt reactors are used for leading power factor circuits, as in lightly loaded cables.

Series Capacitors

The major drawback of series capacitors is: high overvoltages are produced when a short circuit current flows through the capacitor and special capacitive devices have to be incorporated (*e.g.*, spark gaps).

Merits of series capacitors

1. If the load Var requirement is small, series capacitors are of little use.
2. If voltage drop is the limiting factor, series capacitors are effective. Voltage fluctuations due to arc furnaces are also evened out.
3. If the total line reactance is high, series capacitors are very effective and stability is improved.

Synchronous Compensators

Synchronous motors running without mechanical load can absorb or generate Q depending on the excitation. As the synchronous motor losses are considerable compared to static capacitors, the power factor is not zero. When used with a voltage regulator, the compensator can automatically run overexcited at times of high load and underexcited at light loads. The advantage of the synchronous compensator is its flexibility of operation for all load conditions.

Reactive Power Requirement for the Voltage Control of Long Lines

An expression for determining the Q at the receiving end may be derived in terms of the receiving end power Pr, complex A, B parameters of the line such that the receiving end voltage, Vr is equal to or a specified ratio of the sending end voltage, Vs.

15.1.3 Tap Changing Transformers

To determine the tap changing ratios required to completely compensate for the voltage drop in the line, the product of the transformer tap ratios ts and tr at the two ends of the line is made equal to unity. This ensures that the overall voltage level remains in the same order and the minimum range of taps in both transformers is used.

OBJECTIVE TYPE QUESTIONS

1. Reactive power in a single-phase a.c. circuit is a product of
   (*a*) voltage times the current in a circuit.
   (*b*) voltage times the current times the cosine of the angle between the voltage & current.
   (*c*) voltage times current times the sine of the angle between the voltage & current.
   Ans. (*c*)
2. In a lightly loaded transmission line of such a length that the capacitive reactance is appreciable, the receiving end voltage
   (*a*) is always less than the sending end voltage.
   (*b*) may exceed the sending end voltage.

(c) is always equal to the sending end voltage.

Ans. (b)

3. On a long high voltage transmission line under heavy load conditions, Var compensation can be provided by installing
   (a) Series inductive reactors
   (b) Series capacitors (c) Shunt inductive reactors

   Ans. (b)
4. If a synchronous motor (properly synchronized to the supply) is running on no load with negligible loss then
   (a) the stator current will be very high
   (b) the stator current will be zero
   (c) the stator current will be very small
   (d) the back emf will be more than the supply voltage

   Ans. (c)
5. The minimum armature current of the synchronous condenser corresponds to
   (a) 0.8 lagging p.f. (b) zero pf lagging.
   (c) 0.8 pf leading. (d) unity pf.

   Ans. (d)
6. When synchronous condensers are used to provide Vars, leading Vars are provided by
   (a) Increasing the field current (b) Decreasing the field current
   (c) Increasing the speed of the machine

   Ans. (a)
7. Vars are characterized by the fact that they always flow
   (a) From points of low voltage to points of high voltage
   (b) From points of high voltage to points of low voltage
   (c) Between two points having the same voltage

   Ans. (b)

PROBLEMS WITH SOLUTIONS

Prob.15.2.1 Cost of power factor improvement

A consumer's load of 250 kW has power factor of 0.78 lagging. The power supply charges are 50 cents per kVA of maximum demand per month plus 0.5 cents per kW hr. The cost of power factor improving apparatus is US dollar 10 per kVA, and the rate of interest & depreciation is 10 %. To what extent the power factor should be improved so that the amount saved is a maximum ? Calculate also the kVA capacity of the power factor improving apparatus.

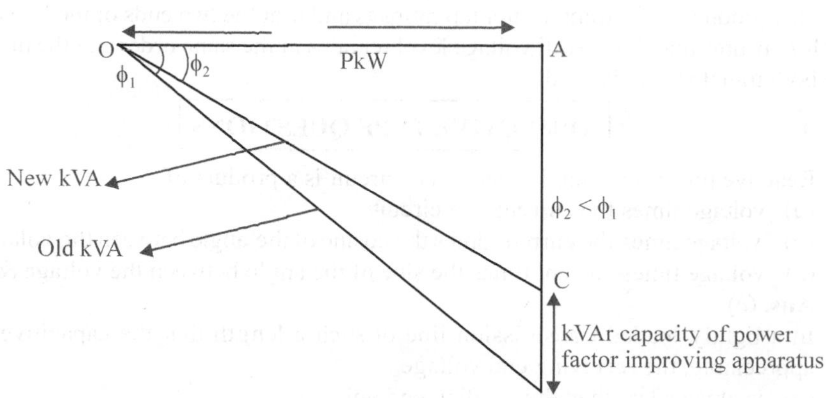

Fig.15.2.1. *P-Q diagram*

pf_1, $\cos\phi_1 = 0.78$ lag

pf_2 , most economical pf, $= \cos\phi_2$

$P = 250$ kW

kVAr capacity of power factor improving apparatus = P ($\tan\phi_1 - \tan\phi_2$)

U	cost of power supply	= 0.5	USD/kVA
E	cost of energy	= 0.05	USD/kW-h
A	cost of apparatus	= 10	USD
I	Interest & depreciation	= 10	%

B Annual charge for power factor improving apparatus $= \dfrac{A \times I \times P(\tan\phi_1 - \tan\phi_2)}{100}$

D Saving by improving the pf is $= U\left(\dfrac{P}{pf_1} - \dfrac{P}{pf_2}\right)$ = Consumer's gross saving

S Consumer's net saving = D – B

To maximize S,

$(dS/d\phi_2) = 0 = (dD/d\phi_2) - (dB/d\phi_2) = -U.P\tan(\phi_2)\times \sec(\Phi_2) + A*(I/100)*P\sec(\phi_2)*\sec(\phi_2)$

or, $\sin\phi_2 = A.(1/100)/U = 0.2$

$\cos\phi_2 = 0.979$ lag **Answer**

Capacity of power factor improving apparatus = $P \times (\tan\phi_1 - \tan\phi_2)$

= 127.623 kVAr **Answer**

Prob. 15.2.2 Reactive Power Required to Re-establish the Fall in Voltage Due to Load

Three supply points A, B and C are connected to a common bus bar M. Supply point A is maintained at a nominal 66 kV and is connected to M through a 66/33 kV transformer (0.1 pu reactance) and a 33 kV line of reactance 20 Ω. Supply point B is nominally at 33 kV and is connected by a 33kV line to M. Supply point C is at 66 kV and is connected to M by a 66/33kV transformer (0.1 pu reactance) and a 33 kV line of 20 Ω reactance.

If at a particular system load, the line voltage at M falls by 5 kV, calculate the magnitude of the reactive volt-ampere injection required at M to re-establish the original voltage.

The pu values are based on a 100 MVA base.

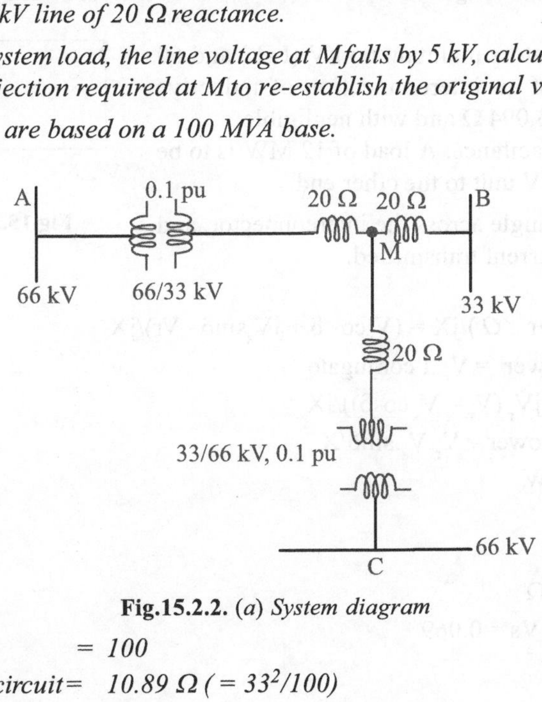

Fig.15.2.2. *(a) System diagram*

MVA_b	*= 100*
X_b in 33 kV circuit	*= 10.89 Ω (= $33^2/100$)*
X_l Line reactance	*= 20 Ω*
$X_l pu$	*= 1.836 pu*

Solution.

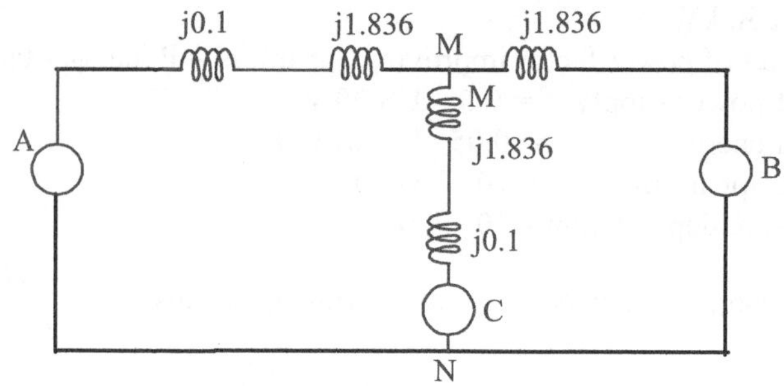

Fig.15.2.2. (b) *Circuit diagram*

X_t Transformer reactance = 0.1 pu
X_a Reactance M to A = 1.936 pu
X_d Reactance M to C = 1.936 pu
Parallel combination of X_a & X_d = 0.968 pu
X_c Reactance M to B = 1.836 pu
Equivalent reactance M to N = 0.634 pu [= 0.968 × 1.836/(0.968 + 1.836)]
Fault MVA at M = MVA_b/equivalent reactance = 157.726 MVA

$$\text{Fault current} = \frac{\text{Fault MVA} \times 1000}{(1.73205 \times 33)} = 2759.501 \text{ A at 0 power factor lagging}$$

dQm/ dVm = 1.73205*fault current = 4779.594 Vars/Volt = 4.779 MVAr /kV
Drop in voltage at M (L-L) = 5 kV
Reactive power required = 23.89 MVAr **Answer**

Prob.15.2.3 Load angle for a given load across an interconnector with the ends maintained at different voltages

Two units of generators maintain 66 kV (L-L) and 60 kV (L-L) at the ends of an interconnection of inductive reactance per phase of 23.094 Ω and with negligible resistance and shunt capacitance. A load of 12 MW is to be transferred from the 66 kV unit to the other end.

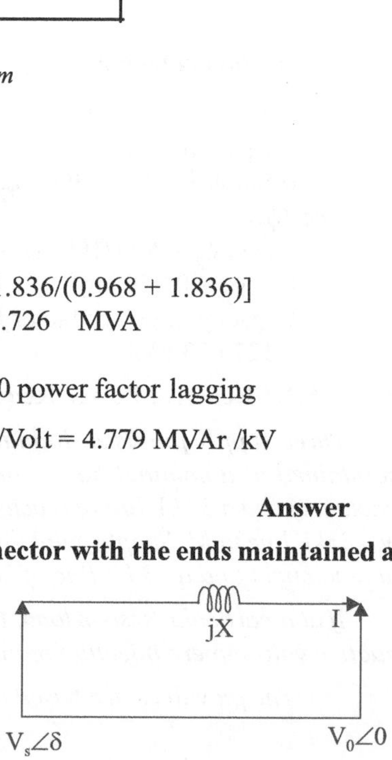

Fig.15.2.3. *Circuit diagram*

Calculate the load angle across the interconnector, and the power factor of the current transmitted.

Solution.

I = ($V_s \angle \delta$ – Vr $\angle O$)/jX = ($V_s$cos δ + j$V_s$sinδ –Vr)/jX
S Complex power = V_r.I conjugate
= ($V_r V_s$sin δ – jV_r(V_r – $V_s$cosδ))/jX
P Real power = V_r.V_s.sinδ/X
P = 12 MW
V_s = 66 kV
V_r = 60 kV
X = 23.04 Ω
sin δ = PX/V_r.Vs = 0.069
δ = 4.003° **Answer**
cos δ = 0.997

$\tan\phi = (-V_s \cos\delta + V_r)/V_s \sin\delta = -1.267$

$\cos\phi = 0.619$ lag **Answer**

Prob.15.2.4 Two generating stations connected by an interconnector- voltage adjustments required to equalize the loads at the two stations

Two generating stations A and B are linked by a line and two transformers of total 20 Ω referred to 132 kV and negligible resistance. A load of 100 MW 0.9 pf lagging is at the bus bars of A and 200 MW, 0.85 p.f. lagging is at the busbars of B. Determine the phase angles between the bus bars of A and B, and the voltage adjustment required to equalize the load on each station. Initially both stations have bus bar voltages of 11 kV which are in phase.

Solution.

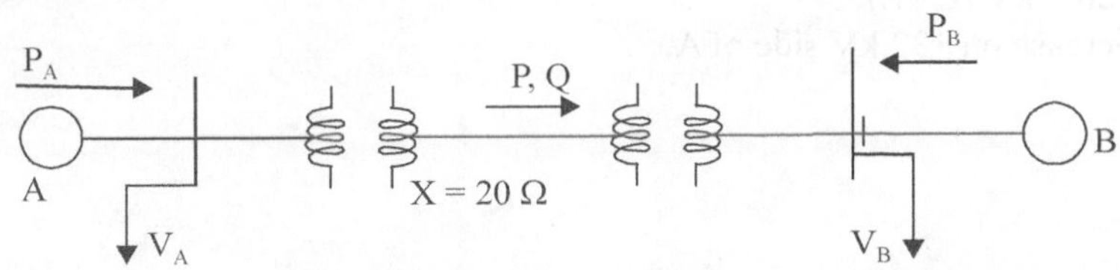

Fig.15.2.4. (*a*) *System diagram*

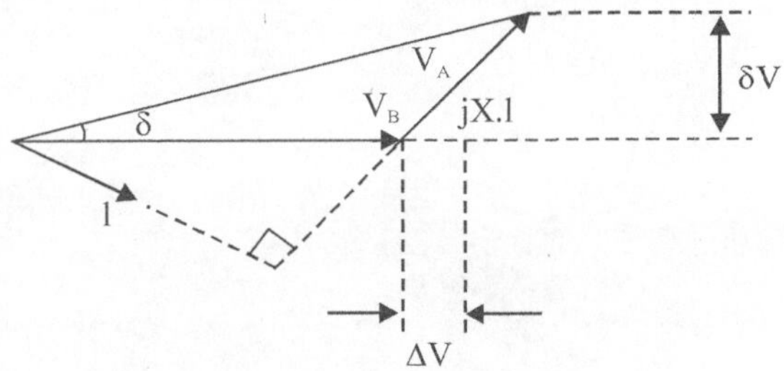

Fig.15.2.4. (*b*) *Phasor diagram*

$P_1 = 100$ MW

p.f.$_1 = 0.9$ lag

$Q_1 = P_1.\tan\left(\cos^{-1}(pf_1)\right) = 48.432$ MVAr

$P_2 = 200$ MW

p.f.$_2 = 0.85$ lag

$Q_2 = P_2.\tan^{-1}\left[\cos^{-1}(pf_2)\right] = 123.948$ MVAr

To equalize the load on each station,

$P_A = P_B = 150$ MW

$P = P_A - P_1 = 50$ MW

$Q_A = Q_B = (Q_1 + Q_2)/2 = 86.190$ MVAr

$Q = Q_A - Q_1 = 37.758$ MVAr, ignoring $Q_{loss} = I^2X$ in line

From the phasor diagram,

$\delta V = X \times P_{ph}/V_B$

$X = 20\ \Omega$

V_B per phase = 76.210 kV

P_{ph} per phase = 16.666 MW (= P/3)

$\delta V = 4.373$ kV

From the phasor diagram,

$\sin\delta = \delta V/V_A = \delta V/V_B = 0.057$

$\delta = 3.291°$ g **Answer**

Hence, voltage of bus bar A is 3.29° in advance of the voltage of bus bar B.

$\Delta V = XQ_{ph}/VA_{ph}$

$Q_{ph} = 12.586$ MVAr

V_A per phase = 76.210 kV

Therefore,

$\Delta V = X\,Q_{ph}/V_{Aph} = 3.302$ kV per ph

$= 5.720$ kV (L– L) **Answer**

i.e., 5.75 kV increase on 132 kV side of A.

CHAPTER 16

ECONOMIC OPERATION

16.1 INTRODUCTION

Look up the site http://www.elec.kitami-it.ac.jp/ueda/java/ELD/default.html for Economic Loading Dispatch Demonstrator which is a Java based demonstration of economic dispatch with three generators on a lossless system.

16.1.1 Formulation of the problem

- Fuel cost and incremental coat curves, concept of incremental fuel rate, Economic Dispatch problem with line losses neglected, no constraints on the generation capacities.
- Economic dispatch problem considering line losses, loss formula and penalty factor, Physical interpretation of coordination equations.
- Simple introduction to the optimal power flow problem.

16.1.2 Economic Dispatch

The module E contained in the notes on Economic operation in the website: **http://cindulkar.tripod.com/index.htm** provides students with visual displays of the matrix approach to solve the economic dispatch problem with the Newton method. The graph of each unit's I/O curve is shown as the solution changes for a given demand level and for unit high and low limits. The student is shown the equations as the solution progresses and is instructed as to what equations have to be added and how to change the parameters for additional analysis.

COST CURVE COEFFICIENTS			
	UNIT 1	UNIT 2	
Quadratic	0.0156	0.0185	K/MWsq-h
Linear	2.1875	2.4074	K/MWh
Constant	120.31	740.74	K

GENERATION RANGE			
	MINIMUM	MAXIMUM	
UNIT 1	200	350	MW
UNIT 2	100	200	MW

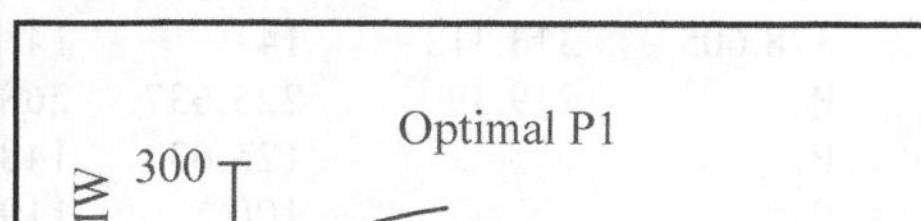

Fig.16.1.2(a)

POWER DEMAND 400 MW

OPTIMAL SOLUTION

P_1 179.766

P_2 220.766

Lambda 7.796

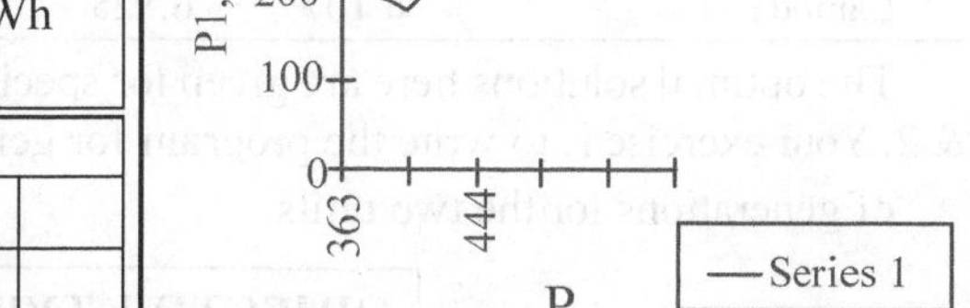

Fig.16.1.2(b)

Notes:

Cost curves

$C_1 \; = A_1 + B_1P_1 + C_1P_1{}^2$

$C_2 \; = A_2 + B_2P_2 + C_2P_2$

$IC_1 \qquad = B_1 + 2C_1P_1$

$IC_2 \qquad = B_2 + 2C_2P_2$

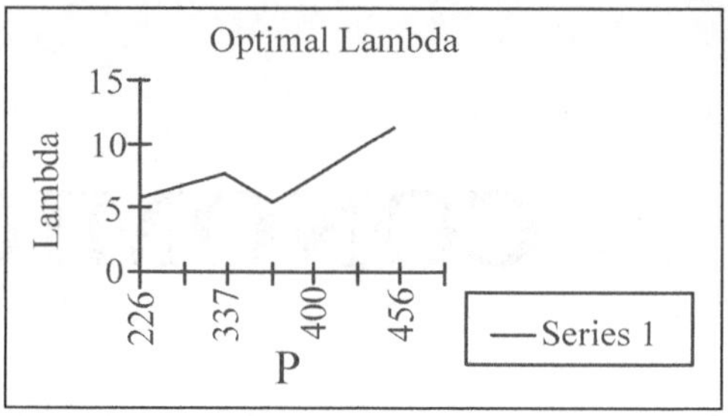

Fig.16.1.2(c)

Optimal dispatch rule and the equality constraint give

$B_1 + 2C_1P_1 = B_2 + 2C_2P_2$

$P_1 + P_2 = P_d$

Solve these equations for P_1 & P_2:

$$P_1 = \frac{C_2P_d}{C_1 + C_2} + \frac{B_2 - B_1}{2(C_1 + C_2)} = 220.233$$

$P_2 = P_d - P_1 = 179.766$

Lambda

P_1	P_2	Lambda_1	Lambda_2	P				
100	119.394	6.825	6.825	219.394				
125.637	100		6.107	225.637	$P_1 = (IC_2 - B_1)/2C_1$			
148.637	119.394		6.825	268.032	$P_2 = (IC_2 - B_2)/2C_2$			
186.298	151.151	8	8	337.449				
200	162.705	8.427	8.427	362.705	2.450		8.427	371.132
179.766	220.766		7.796	400	Given Demand			
244.227	200	9.807	9.807	444.227	$IC_2 = B_2 + 2C_2P_2$, $P_1 = IC_2 - B_1/2C_1$			
250.400	205.205	10	10	455.606				
314.503	259.259	12	12	573.762				
350	289.191	13.107	13.107	639.191				
378.605	313.313	14	14	691.919				
P	219.39	225.637	268.032	337.449	362.7	400	444.227	455.606
P_1		125.63	148.63	186.3	200	179.766	244.227	250.4
P_2		100	119.39	151.15	162.71	220.766	200	205.2
Lambda		6.107	6.825	8	5.912	7.796	10	12

The optimal solutions here are given for specific minimum and maximum generations of units 1 & 2. Your exercise is to write the program for general minimum and maximum values of generations for the two units

OBJECTIVE TYPE QUESTIONS

1. When generating units in a power plant are loaded with equal incremental costs, the cost of generation is
(*a*) maximum
(*b*) minimum
(*c*) not affected
Ans. (*b*)

2. Incremental heat - rate curves for thermal generating units are used to determine the
(*a*) fuel cost in Rs./hour
(*b*) values to which the units should be loaded to decrease the fuel costs to a minimum
(*c*) none of the above.
Ans. (*b*)

3. In economic operation of a power system, the effect of increased penalty factor between a generating plant and system load centre is
(*a*) to decrease the load on the generating plant
(*b*) to increase the load on the plant
(*c*) to hold the plant load constant.
Ans. (*a*)

PROBLEMS WITH SOLUTIONS

Prob.16.2.1 Load sharing of two generating units – given the incremental fuel costs

The incremental fuel costs (in $ per MW-hr) for two generating units are given by

$$\frac{dC_1}{dP_{g1}} = 0.008\,P_{g1} + 8$$

$$\frac{dC_2}{dP_{g2}} = 0.009\,P_{g2} + 6$$

Determine how a load of 1000 MW should be shared by the two generating units for minimum fuel cost.

Solution. Equate the incremental fuel costs;

$0.008\,P_{g1} + 8 = 0.009\,P_{g2} + 6$ . . . (1)

$P_{g1} + P_{g2} = 1000$ . . . (2)

Solving these two equations we get:

$$\begin{pmatrix} P_{g1} \\ P_{g2} \end{pmatrix} = \begin{pmatrix} 0.008 & -0.009 \\ 1 & 1 \end{pmatrix}^{-1} \begin{pmatrix} -2 \\ 1000 \end{pmatrix}$$

Let

A = 0.008
B = – 0.009
C = 1
D = 1
K = AD – BC = 0.017

$$\begin{pmatrix} P_{g1} \\ P_{g2} \end{pmatrix} = \begin{pmatrix} D/K & -B/K \\ -C/K & A/K \end{pmatrix} \begin{pmatrix} -2 \\ 1000 \end{pmatrix}$$

P_{g1} = 411.764 MW **Answer**

P_{g2} = 588.235 MW **Answer**

Prob.16.2.2 Penalty Factors of Plant

In a 2-plant system, the entire load is located at plant 2, which is connected to plant 1 by a transmission line. Plant 1 supplies 100 MW of power with a corresponding transmission loss of 5 MW. Calculate the penalty factors for the two plants.

Solution.

Since, the whole load is at Plant 2, varying P_2 does not affect the transmission loss, P_{loss}.

Thus, from $P_{loss} = P_1^2 B_{11} + 2P_1P_2B_{12} + P_2^2 B_{22}$, we get

$P_{loss} = P_1P_1B_{11}$ or

$B_{11} = P_1P_1/P_{loss}$

$P_{loss} = 5$ MW

$P_1 = 100$ MW

$B_{11} = 0.0005$ (1/MW)

$dP_{loss}/dP_1 = 2P_1B_{11} = 0.1$

Penalty factor for the first unit, $L_1 = \dfrac{1}{1-\dfrac{dP_{loss}}{dP_1}} = 1.111$ **Answer**

dP loss/$dP_2 = 0$

Penalty factor for the second unit, $L_2 = \dfrac{1}{1-\dfrac{dP_{loss}}{dP_2}} = 1$ **Answer**

Prob.16.2.3 Load Sharing of two Units Considering Penalty Factors

For the system of Prob.16.2.2 λ = \$15/MWh and the incremental fuel costs for the two plants are given by :

$dF_1/dP_1 = 0.01\ P_1 + 10$ *(\$/MW hr)*

$dF_2/dP_2 = 0.02\ P_2 + 12$ *(\$/MW hr)*

How much power should be generated at each plant for minimum total fuel cost?

Solution. The penalty factors from Prob.16.2 .2 are:

$L_1 = 1.111$

$L_2 = 1$

$\lambda = 15$ \$/MWhr

$dF_1/dP_1 = \lambda * L_1$

or, $(0.01P_1+10)*L_1 = l$

or, $P_1 = ((\lambda/L_1) - 10)/.01 = 350.135$ MW **Answer**

Similarly,

$dF_2/dP_2 = \lambda * L_2$

or, $(0.02P_2+12)*L_1 = \lambda$

or, $P_2 = ((\lambda/L_2) - 12)/.02 = 150$ MW **Answer**

Prob.16.2.4 Saving in operating costs for economic load sharing compared to equal load sharing between two units

Incremental fuel costs for two generating units in a plant are given by

$dF_1/dP_1 = .02\ P_1 + 4$ *(\$/MW – hr)*

$dF_2/dP_2 = .024\ P_2 + 3.2$ *(\$/MW – hr)*

Total load varies from 50 to 25 MW.

Maximum and minimum loads on each unit are to be 125 and 20 MW respectively.

Determine the saving in fuel cost in \$ /hour for economic distribution of a total load of 180 MW, compared with equal distribution.

Solution. Equating incremental costs, we get

$0.02\ P_1 - 0.024\ P_2 = -0.8$ . . . (1)

$P_1 + P_2 = 180$ . . . (2)

Rewrite:

$0.02\ P_1 - 0.024\ P_2 = -0.8$

$0.024P_1 + 0.024P_2 = 4.32$

Add:

$0.044 \times P_1 = 3.52$

$P_1 = 80$ MW **Answer**

$P_2 = 100$ MW **Answer**

Increase in cost for Unit 1:

$$\int_{80}^{90}(0.02P_1+4)\,dP_1=0.02\times 90\times\frac{90}{2}+4\times 90-0.02\times 80\times\frac{80}{2}-4\times 80$$

for Unit 2:

$$\int_{100}^{90}(0.024P_2+3.2)dP_2=\frac{0.024\times 90\times 90}{2}+3.2\times 90-\left(\frac{0.24\times 100\times 100}{2}+3.2\times 100\right)$$

$= -54.8$ \$/hr Decrease in cost

Net increase in cost = 2.2 \$/hr **Answer**

Prob.16.2.5 Load sharing of two generating units -given the variable operating costs

The variable operating costs of two generating units are given by

$C_1= 10P_1 +.008P_1*P_1$ *\$/hr*

$C_2= 8P_2+ 0.009P_2*P_2$ *\$/hr*

If the total load demand varies from 120 to 140 MW, determine the power output of each unit(coal-fired), the incremental operation cost, and the total operating cost CT that minimizes CT for the load demands, 120,130, and 140 MW.

Solution. For minimum cost, $dC_1/dP_1= dC_2/dP_2$,

$dC_1/dP_1=$ a + bP_1, and

$dC_2/dP_2=$ c + d*P_2, where

a	=	10	10	10
b	=	0.016	0.016	0.016
c	=	8	8	8
d	=	0.018	0.018	0.018
P_T, total load, MW		**120**	**130**	**140**

For minimum cost, $a + bP_1= c + dP_2$........ eq. 1

$P_2= P_T- P_1$............................... ... eq. 2

Using (2) in (1),

$P_1=$ (d*P_T+ c – a)/(b + d) 4.705 10 15.294 MW **Answer**

$P_2= P_T- P_1$ 115.294 120 124.705 MW **Answer**

Incremental cost when C_T is minimized is

$dC_2/dP_2= dC_1/dP_1= a + bP_1=$

10.075 10.16 10.244 \$/MWh **Answer**

Minimum total operating cost is $C_T= C_1+ C_2= aP_1+ bP_1*P_1+ cP_2+ dP_2*P_2$

1209.03 1320.8 1434.2 \$/h **Answer**

Prob.16.2.6 Load sharing of two generating units-given the variable operating costs-considering losses

Total transmission loss for the power system of Prob.16.2.5 is given by

$P_L = .00015P_1*P_1 +.00002\, P_1*P_2 +.00003P_2*P_2$ *(MW)*

P_1,P_2 *are in MW*

Determine the output of each unit, total transmission loss, total load, total cost, when the area λ = *10.1 \$/MWh*

Solution. Let $P_L = eP_1+ fP_1*P_2+ gP_2* P_2$, where

e = 0.00015

f = 0.00002

g = 0.00003

λ = 10.1 \$/MWh

$\lambda = dC_1/dP_1 = (dC1/dP_1)*(1/(1-dP_L/dP_1)) = (a + bP_1)/(1 + 2eP_1 - fP_2)$

$\lambda = dC_2/dP_2 = (dC_2/dP_2)*(1/(1-dP_L/dP_2)) = (c + dP_2)/(1 + 2gP_2 - fP_1)$

Rearranging the above two equations:

$AP_1 + BP_2 = X$ . . . 1

$CP_1 + DP_2 = Y$, where . . . 2

A $= b + 2e\lambda = 0.0190$

B $= f\lambda = 0.00020$

C $= f\lambda = 0.00020$

D $= d + 2g\lambda = 0.01860$

X $\lambda - a = 0.1$

Y $\lambda - c = 2.1$

Solving eqs.1 &2, we get

$P_1 = (DX - BY)/(AD - BC) = 4.05726$ MW **Answer**

$P_2 = \dfrac{-CX + AY}{AD - BC} = 112.822$ MW **Answer**

Loss, $P_L = eP_1^2\ fP_1P_2 + gP_2 = 0.393$ MW **Answer**

$P_T = P_1 + P_2 - P_L = 116.486$ MW **Answer**

$C_T = C_1 + C_2 = aP_1 + bP_1P_1 + cP_2 + dP_2P_2 = 1187.66$ \$/h **Answer**

Prob.16.2.7 Economic choice of plant

A load of 75 kW is to be supplied for 2500 h/year by a motor and the overall power factor must be not less than 0.9. A new induction motor of efficiency 92% and power factor 0.9 and a second -hand motor of efficiency 91% and power factor 0.83 are available. Lossless capacitors costing IR 700 per kVA can be used to raise the power factor of the second-hand motor to 0.9. All interest and depreciation charges may be taken as 12% per year and the tariff for energy is IR 600 per kVA of maximum demand plus IR 45 per kWh. By how much should the price of the second-hand motor be less than that of the new motor in order to make its use economical ?

Solution.

New motor:

Output = 75 kW
Efficiency = 0.92
Input = Output/Efficiency = 81.52 kW
Loss = Input-Output = 6.52 kW
Cost per kWh = 45.00 IR
Hours = 2500.00
Cost of loss = Loss × Hours × cost /Wh = 7337
power factor = 0.9
Maximum demand = output/(efficiency × power factor) = 90.6 kVA
Tariff for energy = 600 IR
Cost = 54348 (= 90.6 × 600)
Total yearly cost = 61685 IR

Second-hand motor:

Output = 75 kW
Efficiency = 0.91
Input = 82.42 kW
Loss = Input-Output = 7.42 kW
Cost per kWh = 45.00 IR

Hours = 2500.00

Cost of loss = Loss* hours* cost per Wh = 8345 IR

Raised power factor = 0.9

$$\text{Maximum demand} = \frac{output}{Efficiency \times Power\ Factor} = 91.6\ \text{kVA (with capacitors)}$$

Tariff for energy = 600 IR

cost = 54945 IR

original pf of second-hand motor = 0.83

$$\text{Rating of capacitors} = \text{output}\left[\tan\cos^{-1}(p.f._{original} - \tan\cos^{-1}(p.f.raised)\right] = 14.1\ \text{kVAr}$$

capacitor cost = 700.0 per kVA

Cost = 9853.304 IR

Interest & Depreciation = 12% =0.12

yearly cost = 1182.396 IR

Total yearly cost = 64472 IR

Yearly difference = 2787 IR

Difference in cost of second-hand motor = yearly difference/Interest & depreciation

= 23229 IR **Answer**

CHAPTER 17

PER UNIT SYSTEM

17.1 INTRODUCTION

17.1.1 Definition & importance

Per unit (pu) system is equivalent to normalizing the quantities use and has special importance in the area of power system and machines, which involve the use of transformers. The per unit system values can also be expressed as per cent values.

pu value = actual value in any unit/ base or reference value in the same unit

17.1.2 How to choose base or reference values?

For a single machine or apparatus, the base value chosen is often the rated values of the machine and therefore in pu system their values become 1pu each. However, when several machine or transmission lines are involved the common value chosen should be such that the calculated pu values should not be too large or too small compared to unity. This helps in improving numerical accuracy in many involved computations.

17.1.3 How many independent base quantities could there be in a power system?

They could be two. Usually, base VA (or MVA_b), and base voltage (or kV_b).
Others are derived quantities, such as,
I_b, Z_b, R_b, X_b, P_b etc.

17.1.4 Base quantities in three- phase circuits

The numerical values of base voltage or base current may differ depending upon whether it refers to phase quantities or line quantities, but pu quantities remain unchanged whether expressed on line to line basis or phase basis.

For three -phase circuit s, one may define

$I_b = VA_b/ (\sqrt{3}\ V_b)$

Where VA_b is three-phase value, V_b is line value and I_b is line current.

But base impedance is always defined on a per phase basis:

$Z_b + V_b\ (L\text{-}L) / (\sqrt{3}\ I_b) = -$ Vph/Iph

$= V_b^2/ VA_b =$ (base line voltage) ^{2}/ base VA

= (Base line voltage in kV) ^{2}/base MVA

17.1.5 Change of base values

Often the impedances of the transformers are calculated on their own ratings i.e. they have a pu value based on their own ratings and then may be required to be changed to another base value. In such cases,

Zpu (new) = Zpu (old) V_b (old)^2 VA_b (new)/ [V_b (new)^2 VA_b (old)]

Prove the above expression.

17.1.6 Advantages of per unit system

- √3 factor is eliminated
- pu voltage drops, reactance etc for same type of design are nearly the same irrespective of ratings etc

17.1.7 Inclusion of transformers in transmission circuit

1. Actual transformer circuit with primary and secondary circuit leakage. Review of its equivalent circuit connected through ideal transformer.
2. Nominal turns ratio & tapping
3. Base voltages & base currents through ideal transformers. Base voltage in transmission line vis-à-vis other parts should be stressed.
4. Effect of phase shift in transformers for voltage & current. For positive, negative, and zero sequences.
5. It may be stressed that a transformer need not work at its rated value and the single- line diagram of a power system consisting of a generator, load, transmission line, and the sending end and receiving end transformers could be converted first to a pu impedance diagram and then the calculations made.

- What happens to the pu quantities if one of the transformer say at the sending end is at + 5% tap value ?
- What happens if one of the transformers is star-star connected and the other is star-delta connected in the same system ?
- Should the base power be expressed in MVA or MW? What happens to the power factor?
- Study typical pu values of reactances of turbo- and hydro- generators ?

OBJECTIVE TYPE QUESTIONS

1. The synchronous reactance of a modern large size turbo-generator in pu is nearest to
   (*a*) 1
   (*b*) 2
   (*c*) 3
   Ands: (*b*)
2. The base value of a power system is chosen based upon
   (*a*) the rating of one machine in the system
   (*b*) the largest machine in the system
   (*c*) some average (rounded) values such that the pu machine ratings are not too large or too small.
   Ans. (*c*)
3. A bank of three single-phase transformers is star-star connected and has a % reactance of 10%. The same bank is then converted to satisfy rated star-delta connections. The per cent reactance of the transformer will now be
   (*a*) $10/\sqrt{3}$ %
   (*b*) $10\sqrt{3}$ % (*c*) 10 %
   Ans. (*c*)
4. A transmission line has a pu reactance of 30 %. If the working voltage is now increased to 110% of its original voltage (the MVA rating of the line remaining the same), the pu reactance of the line will now be
   (*a*) 33%
   (*b*) 36.3% (*c*) 24.8%
   Ans. (*c*)

5. The percentage reactance can be converted into ohmic value with the following formula
   (*a*) X(Ω) = X(%)*100*kV^2 /kVA
   (*b*) X(Ω) = X(%)*1000*kV^2 /kVA
   (*c*) X(Ω) = X(%)*kV^2 /(1000*kVA)

Ans. None of the above

PROBLEMS WITH SOLUTIONS

Prob.17.2.1

A portion of a power system consists of two generators in parallel connected to a step-up transformer that links them to a 230 kV transmission line. The ratings of these components are:

Generator 1: 10 MVA, X = 10%
Generator 2: 5MVA, X = 8%
Transformer: 15 MVA, X = 6%
Transmission line: 230 kV, 4 + j 60 Ohms
Express the reactances and the impedance in % with 15 MVA as the base value.

Solution: Generator 1:

X% on generator rating = 10
MVA rating of generator = 10
Common base MVA selected = 15
X% on common base = (X% on generator rating*Common base selected)/MVA rating of generator = 15 % **Answer**

Generator 2:
X% on generator rating = 8
MVA rating of generator =5
Common base MVA selected = 15
X% on common base = X% on generator rating*Common base selected/MVA rating of generator
= 24 % **Answer**

Transformer :
X% on transformer rating = 6
MVA rating of transformer = 15
Common base MVA selected = 15
X% on common base = X% on transformer rating*Common base selected/
MVA rating of transformer = 6 % **Answer**

Transmission Line:
Line R = 4 ohms
Base kV = 230 kV
Line % R = Line R in Ohms*Base MVA*100/ ((base kV)*(base kV))
= 0.113 **Answer**

Line X = 60 ohms
Base kV = 230 kV
Line % X = Line X in Ohms*Base MVA*100/ ((base kV)*(base kV))
= 1.701 **Answer**

Prob.17.2.2

Draw the pu reactance diagram for the system shown in Fig.17.2.2 (a):

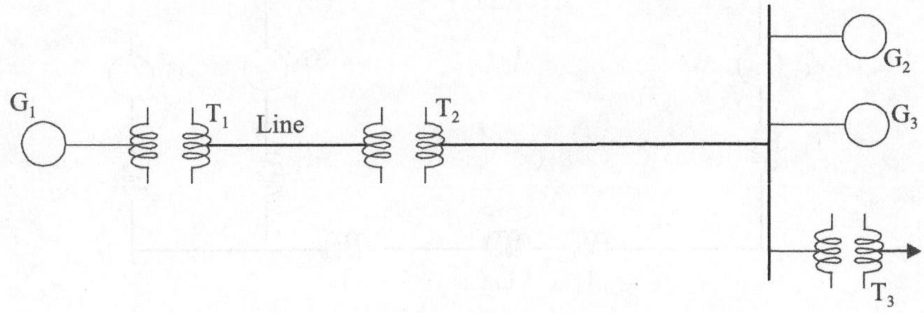

Fig, 17.2.2. *(a) System diagram*

G_1 11kV, 20MVA, 15%
T_1 11/66kV, 30MVA, 15%
Line j60 Ω
T_2 66/11kV, 30MVA, 15%
G_2 11kV, 10MVA, 10%
G_3 11kV, 10MVA, 10%
T_3 11/6.6 kV, 2.5 MVA, 8%
Choose a base of 20MVA, 66kV

Solution: Base MVA 20 MVA

G_1:	X	0.15	pu	on rated MVA
G_2:	X	0.1	pu	on rated MVA
G_3	X	0.1	pu	on rated MVA
G_1:	Rated MVA	20	MVA	
G_2:	Rated MVA	10	MVA	
G_3:	Rated MVA	10	MVA	
T_1:	X	0.15	pu	on rated MVA
T_2:	X	0.15	pu	on rated MVA
T_3:	X	0.08	pu	on rated MVA
T_1:	Rated MVA	30	MVA	
T_2:	Rated MVA	30	MVA	
T_3:	Rated MVA	2.5	MVA	

We now calculate reactances on base MVA=Base MVA*X/rated MVA

G_1:	X	0.15 pu
G_2:	X	0.2
G_3:	X	0.2
T_1:	X	0.1
T_2:	X	0.1
T_3:	X	0.64
Line:	X Ω	60
Base	kV	66

X_{pu} = X Ω × Base MVA/ ((Base kV) × (Base kV))
= 0.275

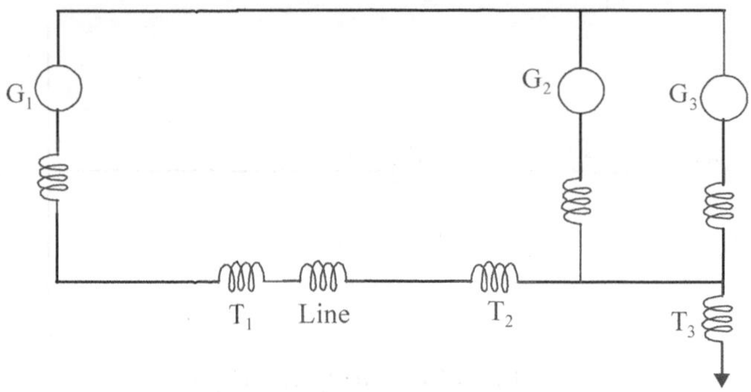

Fig.17.2.2. (*b*) *Reactance diagram.*

Prob.17.2.3

Draw a per unit impedance diagram for the system described below. Choose a base kVA of 50 Two generators G_1 and G_2 are in parallel and are connected to a generator transformer T_1 which, in turn, is connected to a transmission line L. At the receiving end of the line is connected a transformer T_2, the secondary of which is connected to a motor M.

The equipment ratings are given below:

G_1 10kVA.2500V, X = 0.2 pu

G_2 20kVA, 2500V, X = 0.3pu

T_1 40kVA, 2500/8000V, X = 0.1pu

T_2 80kVA, 10000/5000V, X = 0.09pu

Line 50 + j200 Ω

M 25kVA, 4000V, X = 0.1 pu

Solution: Base kVA = 50 kVA

G_1: X	=	0.2	pu	on rated kVA
G_2: X	=	0.3	pu	on rated kVA
G_1: Rated kVA	=	10	kVA	
G_2: Rated kVA	=	20	kVA	
T_1: X	=	0.1	pu	on rated kVA
T_2: X	=	0.09	pu	on rated kVA
T_1: Rated kVA	=	40	kVA	
T_2: Rated kVA	=	80	kVA	
M: X	=	0.1	pu	
M: Rated kVA	=	20	kVA	

We now calculate reactances on base kVA = base kVA*X/rated kVA

G_1: X = 1 pu **Answer**

G_2: X = 0.75 pu **Answer**

T_1: X = 0.125 **Answer**

For T_2, we calculate the reactance on base kVA and base kV

= (Base kVA× X/rated kVA) × (Rated kV/Base kV) × (Rated kV/Base kV)

Rated kV = 10 kV

Base kV = 8 kV

T2: Xpu = 0.087 pu **Answer**

Line: R Ω = 50 Ω

X Ω = 200 Ω

Base kV = 8 kV

Rpu = R Ω × (Base kVA/1000)/ ((Base kV) × (Base kV))

Xpu = X Ω × (Base kVA/1000)/ ((Base kV) × (Base kV))
= 0.156 pu **Answer**

M: Xpu = 0.25 pu **Answer**

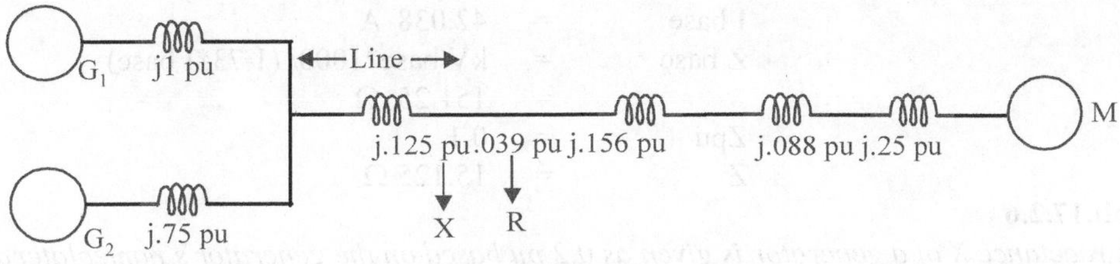

Fig.17.2.3. *Impedance diagram*

Prob.17.2.4

A 3-phase 13 kV transmission line delivers 8 MVA of load .The per phase impedance of the line is .01 +j .05 pu referred to 13 kV, 8 MVA base. What is the voltage drop across the line?

Solution. Base MVA = 8 MVA

Base kV = 13 kV

Base current = base MVA × 1000/ (1.73 × Base kV)
= 355.7137 A

New base MVA/phase	=	2.666	MVA
New Base kV/phase	=	7.514	kV
New Base current	=	355.713	A
R	=	0.01	pu
X	=	0.05	pu

Rpu, new = R × (Base KV/New Base kV/phase) × (Base kV/New Base kV/phase) × (New Base MVA/Base MVA) = 0.0099 pu

Xpu, new =X ×(Base KV/New Base kV/phase) × (Base kV/New Base kV/phase) × (New Base MVA/Base MVA) = 0.049 pu

Base Z new = New Base kV /phase × 1000/New Base current
=21.125 Ω

Actual R = Rpu, new × Base Z new
= 0.210

Actual X = Xpu, new × Base Z new
= 1.053

I × (Actual R) = 74.947 Volts

I × (Actual X) = 374.73 Volts

Voltage drop/phase = 75 + j 375 Volts **Answer**

Prob.17.2.5

A 3-phase item of equipment is rated at 800kVA, 11kV and has an impedance of 10% referred to its voltage.

(*i*) Is 800 kVA understood to be its per-phase or 3phase power rating?

(*ii*) Is 11 kV understood to be its phase-to-neutral or phase-to-phase voltage rating?

(*iii*) What is the impedance of the equipment in ohms?

Solution. (*i*) 3-phase

(*ii*) phase-phase

(*iii*) kVA base = 800

kV base = 11

I base = 42.038 $\left(=\frac{kVA_{base}}{\sqrt{3}\,kV_{base}}\right)$

I base = 42.038 A

Z base = kV base*1000/ (1.73*I base)

= 151.25 Ω

Zpu = 0.1

Z = 15.125 Ω **Answer**

Prob.17.2.6

The reactance X of a generator is given as 0.2 pu based on the generator's nameplaterating of 13.2 kV, 30 MW. The base for calculations is 13.8 kV, 50 MW. Find X on this new base.

Solution: Xpu old = 0.2

base MVA old = 3.0

base MVA new = 50

base kV old = 13.2 kV

base kV new = 13.8 kV

Xpu new = Xpu old*base MVA new × Base KV old × Base kV old/ (base MVA old × base kV new × base kV new)

= 0.304 pu **Answer**

Prob.17.2.7

In the power system shown in Fig.17.2.7 (a):

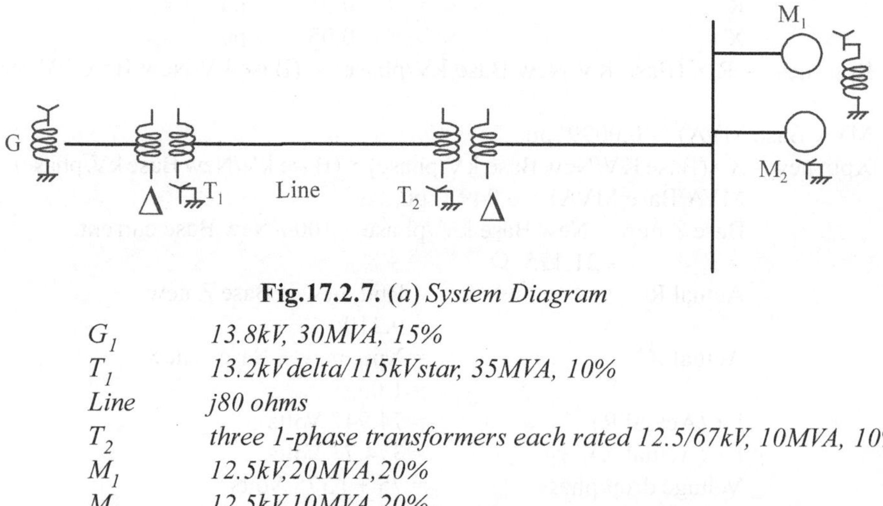

Fig.17.2.7. (*a*) *System Diagram*

G_1 13.8kV, 30MVA, 15%

T_1 13.2kVdelta/115kVstar, 35MVA, 10%

Line j80 ohms

T_2 three 1-phase transformers each rated 12.5/67kV, 10MVA, 10%

M_1 12.5kV,20MVA,20%

M_2 12.5kV,10MVA,20%

Draw the reactance diagram with all reactances marked in pu.

Select the generator rating as base in the generator circuit.

Solution. T_2:

MVA 3-phase = 30 MVA

V_p L-L voltage primary = 12.5 kV

V_s L-L voltage secondary = 115.91 kV

X = 0.1 pu

T_1: = 0.1 pu

MVA 3-phase = 35 MVA

$V_{p'}$ L-L voltage primary = 13.2 kV

V_s' L-L voltage secondary = 115 kV
X = 0.1 pu

A base of 30 MVA ,13.8 kV in the generator circuit requires a 30 MVA base in all parts of the circuit and the following voltage bases.

G_1: kV Base = 13.8 kV
MVA Base = 30 MVA
X = 0.15 pu

Line: =13.8 × 115/13.2
=120.227 kV

Motors: =120 × Vp/Vs
=12.941 kV

Transformer reactances converted to proper base are:

T_1:
X = 0.1 × (30/35) × (13.2 × 13.2/(13.8 × 13.8))
= 0.078 pu

T_2:
X = 0.1 × 12.5 × 12.5/(12.941 × 12.941)
= 0.093 pu

Line:
base X = 120.2273 ×120.2273/30
= 481.819 Ω
X = 80 Ω
X_{pu} = 0.166 pu

M_1:
kV = 12.5
MVA = 20
X own base = 0.2 pu
new base = 0.2 × (30/20) × (12.5 × 12.5/(12.941 × 12.941)
= 0.279 pu

M_2:
kV = 12.5
MVA = 10
X own base = 0.2 pu
new base = 0.2 × (30/10) × (12.5 × 12.5/(12.941\× 12.941)
= 0.559 pu

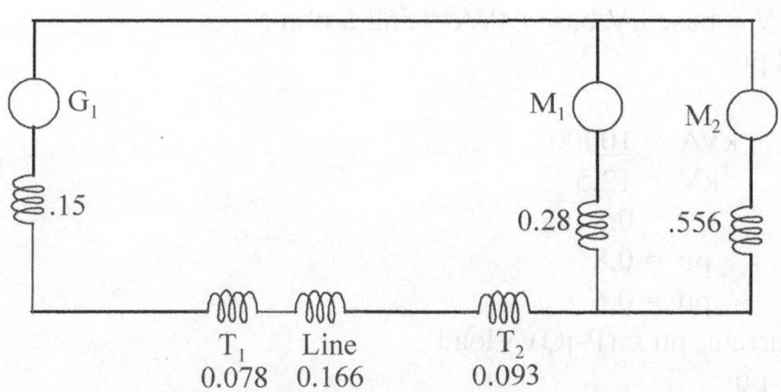

Fig.17.2.7. (b) Reactance diagram: All values in pu

Prob.17.2.8. *A 15000 kVA, 8.5 kV 3-phase generator has a sub-transient reactance of 20 %. It is connected through a delta-star transformer to a high voltage transmission line having a total series reactance of 70 ohms. At the load end of the line is a star-star step-down transformer. Both* *transformer banks are compose of single-phase transformers connected for three-phase operation. Each of the three transformers composing each bank is rated 6667 kVA, 10-100 kV, with a reactance of 10 %. The load represented as impedance draws 10000 kVA at 12.5 kV and .8 pf lagging.*

(*i*) *Draw the one-line diagram and mark base kV in the three parts of the system. Choose a base of 10000 kVA, 12.5 kV in the load circuit*

(*ii*) *Draw the positive sequence impedance diagram showing all impedances in pu.*

(*iii*) *Determine the voltage at the terminals of the generator.*

Solution.

Base kV = 7.22 Base kV=125 base kV=12.5

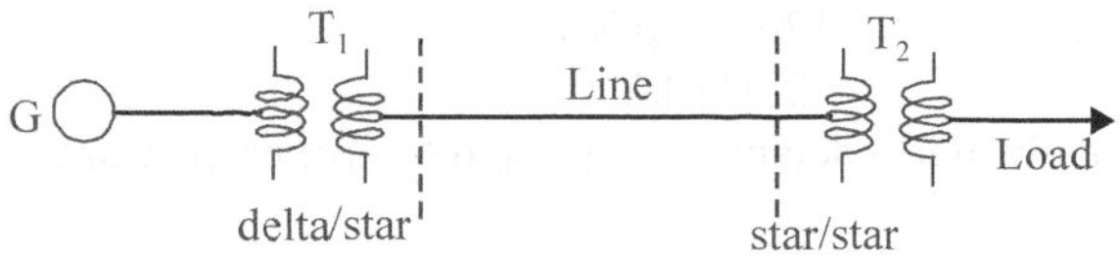

Fig.17.2.8. *(a) System Diagram*

Base kVA = 10000 in load

Base kV = 12.5 in load

T_1 and T_2:

single-phase units

kVA per phase = 6667

kVA 3-phase = 20001

X = 0.1 pu

T Turns Ratio = 0.1

kV(L-L) High side = 125 kV

kV(L-L) Low side = 7.216 kV

kV single-phase-low side = 10 kV

X on common base =.1 × (10000/(3 × 666)) × (10 × 1.73/12.5) × (10 × 1.73/12.5) =0.095 pu

G_1: kVA = 15000

kV = 8.5

X'' = 0.2 pu

Base kV =(125/1.73205) × (T) = 7.216

X'' = on common base = X'' × (base MVA/gen. MVA) × (gen KV/base kV) × (gen kV/base kV) =0.184 pu

Line:

kV = 125

X = 70 Ω

Base Z = base kV × base kV/base MVA=1562.5 ohms

X pu = 0.0448 pu

Load:

kVA = 10000

kV = 12.5

pf = 0.8 lag

P_{load} pu = 0.8

Q_{load} pu = 0.6

I, load current, pu = (P-jQ)/Vload

V_{load} = 1 pu

Re I = P/V_{load} =0.8

Im I = Q/V_{load} = – 0.6

$R_{load} = 0.8$ pu

$X_{load} = 0.6$ pu

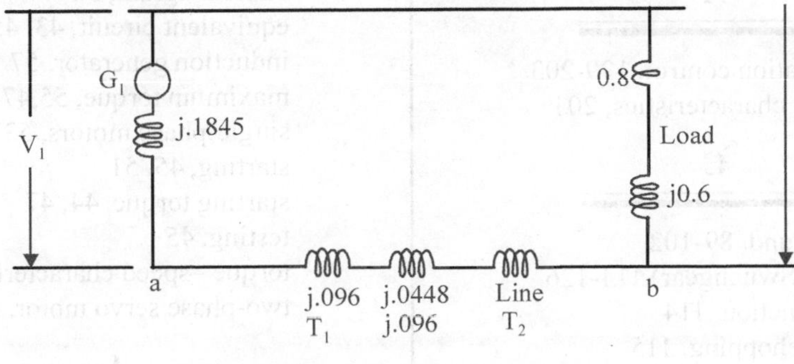

Fig.17.2.8. (*b*) Single- line diagram

Reactance a-b = 0.236 pu

$V_t = 1 + (\text{Re}I + j\text{Im}I) \times (j \times \text{Reactance a-b})$

$\text{Re } V_t = 1 - \text{Im}I \times \text{Reactance a-b} = 1.141$

$\text{Im } V_t = \text{Re } I \times \text{Reactance a-b} = 0.189$

$V_{tmag} = 1.157$pu

$= 8.352$ kV **Answer**

INDEX

A

C

D

E

F

I

L

O

P